AutoCAD

AutoCAD Release 12 for students

A. Yarwood

Copublished in the United States with
John Wiley & Sons, Inc., New York

Longman Scientific & Technical,
Longman Group UK Limited,
Longman House, Burnt Mill, Harlow,
Essex, CM20 2JE, England
and Associated Companies throughout the world.

*Copublished in the United States with
John Wiley & Sons, Inc., 605 Third Avenue, New York, NY 10158*

© Longman Group UK Limited 1994

All rights reserved; no part of this publication may be reproduced, stored in a retrieval system, or transmitted in any form or by any means, electronic, mechanical, photocopying, recording, or otherwise without either the prior written permission of the Publishers or a licence permitting restricted copying in the United Kingdom issued by the Copyright Licensing Agency Ltd, 90 Tottenham Court Road, London, W1P 9HE.

First published 1994

ISBN 0–582–22682–1

British Library Cataloguing in Publication Data

A CIP record for this book is available from the British Library

Library of Congress Cataloging-in-Publication Data

Set in Linotron Melior 10/13 pt. by 8

Produced by Longman Singapore Publishers (Pte) Ltd.
Printed in Singapore

Contents

List of plates	x
Preface	xiii
Acknowledgements	xv

1 Introduction — 1
- Advantages of using a CAD software package — 1
- The AutoCAD Release 12 Workstation — 2
- The IBM compatible PC requirements — 2
- Starting up AutoCAD 12 — 3
- The AutoCAD editor — 5
- The File pull-down menu — 6
- Pull-down menus — 8
- Dialogue boxes — 8
- Selection of commands — 11
- The AutoCAD 12 on-screen menus — 13
- Some necessary settings — 14
- Function key calls — 17
- The use of Ctrl+C — 17
- Revision — 17
- A work disk — 18
- The prototype file *acad.dwg* — 20

2 Constructing simple drawings — 21
- Introduction — 21
- Accuracy in the construction of drawings — 24
- Using the mouse buttons — 25
- The command **LINE** — 25
- The command **DLINE** — 28
- The command **SKETCH** — 31
- The command **ARC** — 31
- The command **CIRCLE** — 33
- The command **ELLIPSE** — 34

	The command **POLYGON**	36
	The command **PLINE (polyline)**	37
	Adding text to drawings	39
	A work sheet	43
	Object snaps	43
	Notes on text and the command **DTEXT**	45
	Exercises	45
3	**Modifying drawings**	**51**
	Introduction	51
	The command **ZOOM**	53
	The Modify menu commands	55
	The command **ERASE**	57
	The command **MOVE**	59
	The command **COPY**	60
	The command **ROTATE**	62
	The command **SCALE**	63
	The command **STRETCH**	65
	The command **TRIM**	66
	The command **BREAK**	68
	The command **EXTEND**	70
	The command **PEDIT**	71
	Revision notes	71
	Exercises	72
4	**Layers**	**76**
	Introduction	76
	The options in the **LAYER** command	76
	The Layer Control dialogue box	77
	The command **CHANGE**	80
	A work disk	81
	An AutoCAD 12 example drawing	83
5	**The Construct pull-down menu**	**85**
	Introduction	85
	The command **MIRROR**	85
	The command **ARRAY**	87
	The command **CHAMFER**	90
	The command **FILLET**	91
	The command **OFFSET**	92
	Hatching	93
	The command **BHATCH**	94
	Exercises (some revision)	100
	A note on text in hatched areas	104

6 The Settings menu. Dimensioning 106
Introduction 106
The Dimension Style dialogue boxes 106
The command **DIM** 110
The action of Modify commands on dimensions 113
Alternative units in dimensions 115
Some rules for AutoCAD 12 dimensioning 116
Exercises 116
Grips 119

7 Orthographic projection 123
Introduction 123
Drawing sheet layout 123
Constructing a First Angle orthographic projection 124
Constructing a Third Angle orthographic projection 128
Sectional views 128
Exercises 128

8 Wblocks, blocks and inserts 135
Introduction 135
The command **WBLOCK** 135
The command **DDINSERT** 137
Blocks 140
Libraries of blocks 141
The construction of circuit drawings from blocks 142
Other examples of libraries 144
Exercises 146

9 Pictorial drawing 149
Introduction 149
The command **SNAP** 149
The command **ISOPLANE** 150
The command **ELLIPSE** 151
The dialogue box **Drawing Aids** 152
Examples of isometric drawings 153
Exercises 154
Isometric drawing is not a three-dimensional method 156
Oblique drawing 157
Planometric drawing 159

10 The 3D Surfaces commands — 160
Introduction — 160
The AutoCAD 3D coordinate system — 160
3D Surfaces commands from the **Draw** pull-down menu — 161
The command **3DFACE** — 162
The command **HIDE** — 163
The command **VPOINT** — 163
The 3D Surfaces commands — 166
The command **EDGESURF** — 166
The command **RULESURF** — 167
The command **REVSURF** — 168
The command **TABSURF** — 168
The command **ELEVATION** — 170

11 The Advanced Modelling Extension (AME) — 172
Introduction — 172
The AME primitives — 172
The AME Boolean operators — 174
Examples of Boolean operations in AME — 176
Examples of AME solid models — 177
The commands **SOLFILL** and **SOLCHAM** — 178
The commands **SOLEXT** and **SOLREV** — 180
Examples of AME solid models — 182
Exercises — 183
Further examples of AME solid models — 185
A more difficult exercise — 187

12 UCS (User Coordinate System) and AME — 189
Introduction — 189
The command **UCS** — 190
Notes on the **UCS** — 192
Some more AME commands — 193
The AME command **SOLCUT** — 194
The AME command **SOLPROF** — 195
The AME command **SOLSECT** — 196
An example of an AME solid model — 198
Exercises — 199
Some notes on the **UCS** and AME — 203

13 Tilemode, MSpace, PSpace and Dview — 204
Introduction — 204
The command **TILEMODE** — 204

	The command **MVIEW**	205
	Viewports	206
	Orthographic projections from AME models	209
	The command **DVIEW**	210
14	**Further exercises**	**213**
	Introduction	213
Appendix A – Orthographic projection		**220**
	Notes on First and Third Angles	224
	Sectional views	224
	Types of line in technical drawings	226
Appendix B – MS-DOS		**227**
	Introduction	227
	Start-up	227
	Some MS-DOS commands	228
Appendix C – Plotting		**233**
Index		**236**

List of plates
(between pages 128 and 129)

Plate I　　The **Hatch Options** dialogue box selected from the **Boundary Hatch** dialogue box, together with **Hatch Patterns** selected from **Hatch Options**. Colours of dialogue boxes set with the aid of the command **DLGCOLOR**.

Plate II　　The **Select Text Style** dialogue box from **Entity Creation Modes**, together with the **Text Style ROMANC Symbol Set** dialogue box from **Select Text Style**. Colours of dialogue boxes set with the aid of the command **DLGCOLOR**.

Plate III　　A 4-viewport screen in **Paper Space** with the pull-down menu **View** showing that **Tilemode** has been set **Off**. The AME solid model was constructed in **Model Space**.

Plate IV　　A 4-viewport screen in **Model Space** showing four views of an AME solid model.

Plate V　　A 256-colour rendering of several AME solid models. Rendering with the AutoCAD Release 12 **AVE Render**.

Plate VI　　A **Vpoint** view of an exploded solid model – a crank from a small compressor – constructed with the aid of AME after the **HIDE** has removed hidden lines.

Plate VII　　The AME solid model of Plate VI after the action of the command **SHADE**.

List of plates

Plate VIII   Selecting a colour for one part of the solid model (Plate VI) from the colour wheel in the **Color** dialogue box.

Plate IX   Checking lighting effects for the colour selected from the colour wheel (Plate VIII) in the **Modify Finish Render** dialogue box.

Plate X   The fully rendered exploded solid model of Plate VI. Rendering in AutoCAD Release 12 with **AVE Render**.

Plate XI   A design for a coupling linkage. AME solid models.

Plate XII   The design of Plate XI rendered with **AVE Render**.

Plate XIII   An AME solid model of a bungalow.

Plate XIV   The bungalow of Plate XIII in a different **Vpoint** position after rendering with the aid of **AVE Render**.

Plate XV   A garden table and chairs constructed with the aid of AME and rendered with the aid of **AVE Render**.

Plate XVI   AutoCAD 386 Release 12 can be run as an **MS-DOS** window in Windows 3.1.

Preface

This book is intended for use by those who wish to learn how to produce technical drawings with the aid of AutoCAD Release 12. AutoCAD is undoubtedly the world's most widely used computer aided design (CAD) software package. More CAD workstations are equipped with AutoCAD in the world than are equipped with all other forms of CAD put together. AutoCAD Release 12 is a considerable update on all previous releases. Its major enhancement is its very modern GUI (Graphic User Interface) method of calling its various facilities. Other enhancements include a considerable increase in the speed of its operation and the ease with which technical drawings can be constructed. It is a *user-friendly* software package. AutoCAD 12 is a very complex package and a book of this size cannot hope to deal with all its complexities. There is enough detail here however to satisfy the requirements of pupils in the upper forms of schools or students in Further and Higher Education. It should also be a valuable text for those in industry new to AutoCAD who wish to learn how to construct technical drawings with the aid of a computer. Because of the complex nature of AutoCAD Release 12, no attempt is made here to describe all its facilities. Among the features not discussed are the variety of set variables controlling the way in which drawings are constructed, the ADS (the AutoCAD Development System), the use of macros to speed up methods of drawing.

Although this book is primarily aimed at those using AutoCAD installed in a PC (Personal Computer), its contents will be as suitable for use by those wishing to learn how to operate the software on those other types of computer for which an AutoCAD package has been designed.

The first chapters are concerned with drawing in two dimensions (2D), later chapters deal with solid model constructions in three dimensions (3D), principally with AME (Advanced Modelling Extension). Details of the excellent rendering feature of Release 12

(AVE Render) are not included, but a number of colour plates show the results of rendering solid models with its aid. In all it is hoped that the reader will be able to develop skills in using AutoCAD Release 12 and become sufficiently intrigued by this remarkable CAD software to take his or her studies further by experimentation with the software and by additional reading among the host of books on AutoCAD which have been written.

The drawings in the book were constructed in AutoCAD Release 12 on a 386 computer fitted with an 80387 math co-processor and 16 megabytes of RAM. The drawings were then plotted to a Roland plotter. The illustrations of dialogue boxes were taken as screen dumps with the aid of the Inset software package Hijack for Windows. The resulting screen dump files were printed on a Hewlett-Packard LaserJet IIIP printer.

Several appendices are included. That dealing with the theory of orthographic projection was included at the request of several lecturers in Further or Higher Education who informed me it would be a useful inclusion for those of their students who had not previously been taught technical drawing. That dealing with MS-DOS is necessarily short because of the limitations of a book of this nature. The final appendix confirms the quality of Release 12 as a user friendly package – plotting and/or printing is now an easy-to-use method of producing hard copy with the aid of the software.

Many of the dialogue boxes which are now an integral part of AutoCAD are described. It is assumed, however, that commands will be entered from the keyboard and that the selection device will be a mouse. The author feels that once a student or beginner can construct in AutoCAD by entering the commands he/she wishes to employ at the keyboard, a better understanding of the structure of the commands system will be gained. Then, other methods such as selecting commands from a pull-down menu, from an on-screen menu or from a graphics tablet overlay, will be that much easier to learn at a later date.

Acknowledgements

The author wishes to acknowledge with grateful thanks the help given to him by members of the staff of Autodesk Ltd.

Registered trademarks

The following are registered in the US Patent and Trademark Office by Autodesk, Inc.:

Autodesk, AutoCAD, AutoShade, AutoSketch, Autodesk 3D Studio, Advanced Modelling Extension (AME).

IBM is a registered trademark of the International Business Machines Corporation.

MS-DOS is a registered trademark of the Microsoft Corporation.

A. Yarwood is a Registered Applications Developer with Autodesk Ltd.

CHAPTER 1

Introduction

Advantages of using a CAD software package

1. A CAD package such as AutoCAD Release 12 can be used to produce any technical drawing which can be produced "by hand";
2. Drawings can be produced much more quickly with CAD than when working "by hand" – speeds of as much as 10 times or more are possible with skilled operators;
3. Drawing with CAD equipment is less tedious than working "by hand" – in particular, the adding of features such as hatch lines and the drawing of notes and other lettering is easier, much quicker and more accurate;
4. Drawings, or parts of drawings, can be copied, scaled, rotated, mirrored or moved with ease. Drawings, or parts of drawings, can be rapidly inserted into other drawings, without having to re-draw the insertion;
5. The same detail need never be drawn twice because it can be copied or inserted with ease. A basic rule when drawing with the aid of CAD is:

    Never draw the same thing twice;

6. New details can be added to a drawing, or detail within a drawing can be altered with ease without having to make any mechanical erasures;
7. Skilled operators can automatically dimension drawings with accuracy, greatly reducing the possibility of dimensional error;
8. Drawings saved to files on disk allows the saving of storage space;
9. Drawings can be plotted and/or printed to any scale required by the user without the need for re-drawing;
10. With AutoCAD Release 12, solid drawing data can be

exchanged with other computer devices such as CAM (Computer Aided Machining);
11. With AutoCAD Release 12, data can be exchanged with other software such as databases and spreadsheets.

The AutoCAD Release 12 Workstation

In this book it is assumed that the computer in which AutoCAD is installed will be an IBM compatible PC (Personal Computer) working under MS-DOS. Some 80 per cent of AutoCAD workstations are equipped with IBM compatible computers. However, it must be realized that AutoCAD packages can be installed on other types of computers and on networks. AutoCAD Release 12 can also be installed as a Windows application if desired. Here we are assuming that the software will be working under MS-DOS 5.0 and, if working in Windows, the Windows package will be version 3.1.

The IBM compatible PC requirements

AutoCAD 12 is not available for a 286 IBM compatible. It is only available for a PC equipped with an 80386 (386) operating system chip, or better – e.g. a 486, or later. If the machine is a 386, a math co-processor (30387) chip is necessary to provide sufficient speed for the computer to work efficiently with AutoCAD. A math co-processor is not necessary with a 486 (or better) system. The addition of a Weitek co-processor chip will increase the speed at which AutoCAD processes commands – if a socket for a Weitek chip is available in the motherboard of the computer. The computer must have at least 4 megabytes of memory and a VGA (Video Graphics Array) screen – or better. Although AutoCAD will work satisfactorily with a monochrome VGA screen, there are many advantages in using a colour monitor.

A mouse is perfectly satisfactory as a pointing/selecting device when working with AutoCAD 12, although a puck or stylus with a graphics tablet is also suitable. If working in Windows 3.1, a mouse will inevitably be the pointing device in use. In this book, it will be assumed that a two-button mouse is the pointing/selecting device of choice. Figure 1.1 shows the workstation set-up on which the information in this book is based.

Selection devices

Although many different types of selection device can be successfully used when working with AutoCAD 12, throughout this book

Introduction 3

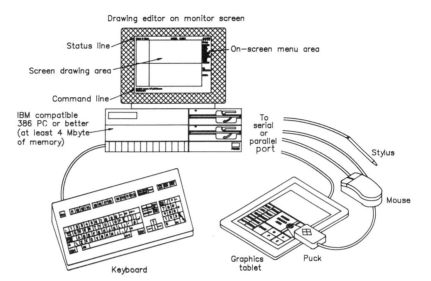

Fig. 1.1 An IBM PC workstation for AutoCAD 12

the selection device of choice will be a two-button mouse. When reference is made to selection with a mouse, the following terms will be used:

Left-click: press the left-hand button and immediately release it;
Right-click: press the right-hand button and immediately release it.

Although in many cases a *right-click* of the mouse produces the same results as pressing the *Return* (*Enter*) key of the keyboard, this is not always so. Thus when the term *Return* is given in this book, the *Return* key of the keyboard must be pressed and not a *right-click* of the mouse. Note that some keyboards have either a *Return* or an *Enter* key. Some keyboards have both keys. Both *Return* and *Enter* keys have the same result.

Note that if using selection devices such as a puck or a stylus with a digitizing tablet or when using a trackerball or joystick, the methods described here can be followed as easily as if using a mouse. This is because these devices will have keys or buttons which produce the same effects and results as the two buttons of a standard mouse.

Starting up AutoCAD 12

Once installed in a PC, AutoCAD 12 may be started up in several ways:

1. By a batch file, e.g. I start up my AutoCAD 12 by typing 12 at

the MS-DOS **C:>** screen prompt on my own PC. This is because my AutoCAD batch file is named *12.bat*. Others may prefer a start-up batch file named *acad12.bat.* or just *acad.bat* – needing either *acad12* or just *acad* to be typed at the keyboard;

2. A method often employed in colleges is for an information screen to be displayed when a PC is switched on giving information to the operator as to the exact keying required to start up AutoCAD. The information might for example show that to bring the college choice of word processor on screen, type *wp*; to bring the college choice of database, type *db*; to bring on introductory AutoCAD lessons, type *acad1* – and so on;
3. In a drawing office, the PC will probably be configured so that when the machine is switched on, AutoCAD 12 will automatically start up.

When AutoCAD 12 starts up, a screen similar to that given in Fig. 1.2 appears, giving information as to the Release number, the date of the Release, the licensee of the software and from where the software was obtained. This screen remains on the monitor for several seconds, then automatically clears and, after the AutoCAD 12 menu utilities have loaded, the AutoCAD drawing editor appears.

Fig. 1.2 The AutoCAD 12 start-up screen

The AutoCAD editor

Figure 1.3 is an example of the AutoCAD drawing editor. Note that although the given illustration shows a white screen surrounded by details in black, the actual colours of the screen depend upon the way in which AutoCAD 12 has been configured. Different colours of screens are shown in the colour plate section starting after page 128. The details on the screen as shown in Fig. 1.3 are those which appear when the drawing command LINE has been selected (or entered at the keyboard). The white area of the screen is that in which details are added to the drawing under construction. The various parts of the drawing editor are:

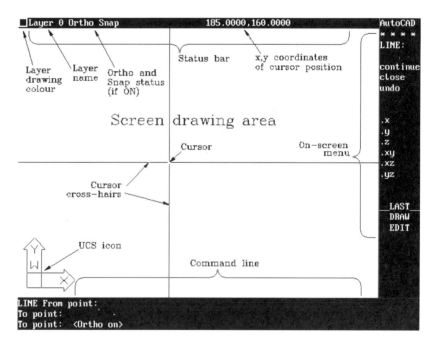

Fig. 1.3 The AutoCAD 12 drawing editor

- *The Screen drawing area*: in which the cursor and the cursor crosshairs can be seen – these can be moved around the screen under the control of the selection device (the mouse). The UCS icon also appears on screen (see page 189);
- *The Status bar*: showing details of the Layer name and its drawing colour, together with the names **Ortho and Snap** – if they are ON (page 15) and the x,y coordinate position of the cursor (page 24);
- *The On-screen menu*: details of the various prompts associated with the command in use appear in this area of the editor. In the

example shown in Fig. 1.3 the prompts associated with the LINE command are shown.

The Command line: the command in action at the time is shown in the command line together with the actions to be taken by the operator (prompts). Usually three lines of information appear at the command line, although AutoCAD can be configured to show less or more lines if thought desirable.

The File pull-down menu

When the drawing editor appears, the operator can commence construction of a new drawing. If a drawing already exists in a file on disk, that drawing can be called back to the screen by moving the cursor into the status bar. A series of pull-down menu names appears in the status bar and a selection arrow replaces the cursor. Select the File menu by moving the arrow under the control of the mouse over the name **File** and *left-click*. The **File** pull-down menu then appears on screen as shown in Fig. 1.4.

Select the command **Open** from the **File** pull-down menu and the menu is replaced by the **Open Drawing** dialogue box. A description of the parts of dialogue boxes is given on page 8. Position the arrow included in this box, under the control of the mouse, over

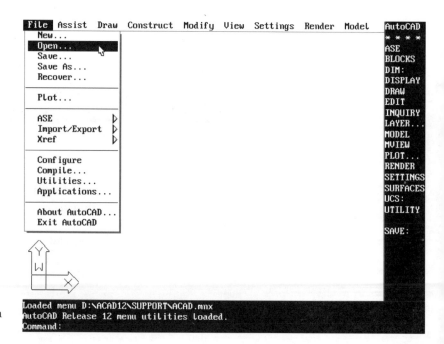

Fig. 1.4 The **File** pull-down menu

Introduction   7

the disk name in which the required file is held. If necessary follow this by selecting the directory in which the file is stored. A list of the AutoCAD drawing files in the chosen directory appears. Select the name of the required drawing file. Finally, *left-click* on **OK** and the chosen drawing appears on screen. An example is given in Fig. 1.5.

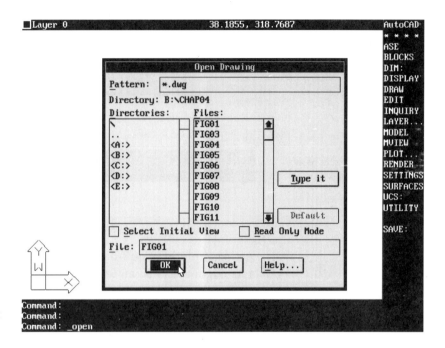

Fig. 1.5 The **Open Drawing** dialogue box

If commencing a new drawing either start its construction immediately after the full drawing editor appears, or select **New** from the **File** pull-down menu. The advantage of starting a new drawing by making this selection is that the drawing can be given a filename before starting its construction. When the time comes to save the drawing to file, it will then be saved with its filename as given by keying a name in the **New Drawing Name** box of the **Create New Drawing** dialogue box (Fig. 1.6).

As can be seen in Fig. 1.4, the **File** pull-down menu includes **Save** and **Save As**. If a name is given to a drawing about to be constructed by using the **Create New Drawing** dialogue box, the file will be saved to that name if **Save** is selected from the pull-down menu. If **Save As** is selected, a name must be keyed at the command line.

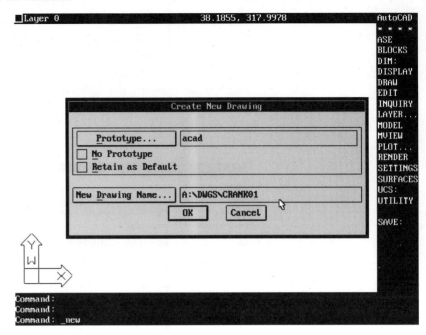

Fig. 1.6 The **Create New Drawing** dialogue box

Pull-down menus

An example of a pull-down menu has already been given in Fig. 1.4. Figure 1.7 is an example of how some pull-down menus produce sub-menus which are *cascaded* on the screen. If a triangular arrow appears against a command name in a pull-down menu, then a sub-menu will appear when that name is selected. The selected sub-menu may itself have sub-menus – indicated again by arrows. In the example given the command name **Dimension** is followed by an arrow, which brings a sub-menu on screen containing the name **Radial**, itself followed by an arrow. A further sub-menu appears when **Radial** is selected.

Note that when a name in a pull-down menu is selected, the name is highlighted.

Dialogue boxes

When three dots follow a command name in a pull-down menu, a dialogue box will appear on screen when that command name is selected.

We have already seen examples of the dialogue boxes which are used with AutoCAD 12 in Figs 1.5 and 1.6. When called, dialogue boxes appear in the centre of the drawing editor. If the selection

Introduction 9

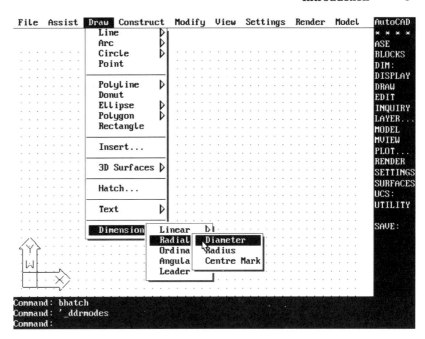

Fig. 1.7 An example of cascading pull-down menus

arrow is moved (under mouse control) into the dialogue box title bar and the left button of the mouse pressed and held down, the dialogue box can be moved under the control of mouse movement. In Fig. 1.8 the **Entity Creation Modes** box has been moved in this

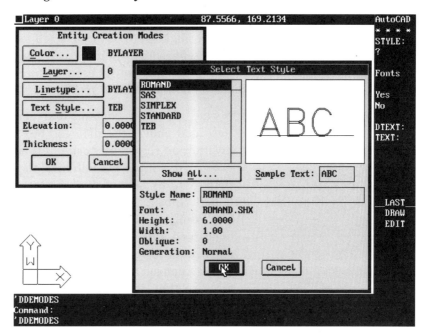

Fig. 1.8 The **Entity Creation Modes** dialogue box

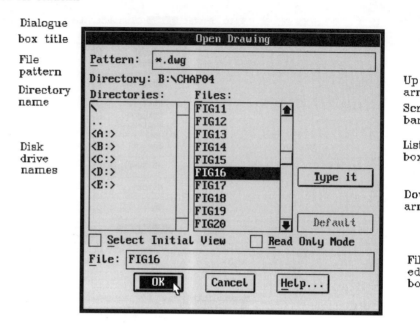

Fig. 1.9 The parts of a dialogue box

manner into the top left-hand corner of the editor. The selection arrow was then placed in the **Text Style** box of the dialogue followed by *left-click*. A second dialogue box then appears – the **Select Text Style** box. This was moved to the bottom right-hand corner. The **Select Text Style** dialogue box includes an *image tile* in which a small area of the dialogue box displays a picture of the letters (or figures) from the selected text style. Other dialogue boxes will be seen which include such image tiles.

The most frequently used features of a typical AutoCAD 12 dialogue box are shown in Fig. 1.9.

Dialogue title bar: contains the name of the dialogue box. As indicated above, the dialogue box can be moved to a different position on screen by positioning the box arrow in the title box, holding down the left button of the mouse followed by dragging the box to its new position before releasing the mouse button;

File pattern box: the file pattern is displayed here in the form of its file extension. Because the dialogue box shown in Fig. 1.9 is concerned with opening drawing files the AutoCAD drawing file extension *.dwg appears in this example. The extension name can be changed by moving the box arrow into the File pattern box area, deleting the given extension name and another extension name keyed in. For example *.dwg could be changed to *.shx or *.sld – two types of file which can be loaded into AutoCAD.

Introduction 11

Disk drive names: position the box arrow over the required disk drive name and *double left-click*. All directories (if any) in the selected disk drive will appear in the disk drive names box. Select the required directory and a list of the drawing files in that directory will appear in the *List box*;

List box: position the box arrow over the required file name in the list box and *left-click*. The selected file name appears in the *File edit box*;

File edit box: a name other than one selected from the *List box* can be keyed into this box by first selecting *Type it* and then keying in the required file name;

Scroll bar: position the box arrow over the square in the scroll bar and press and hold down the left mouse button. The square can then be moved up and down the scroll bar. As this movement takes place the names in the *List box* are scrolled up or down. Alternatively, a *left-click* on the Up or Down arrows moves the names up or down one at a time.

Selection of commands

The construction of drawings in AutoCAD involves a command system – e.g. if one wishes to draw a line, the AutoCAD command which must be called or selected before it can be drawn is LINE. There are four methods by which command names can be called or selected. In each of the following methods the command LINE is taken as a typical example:

1. *By selection from pull-down menus*: position the cursor in the Status bar. The bar changes to show the names of all pull-down menus. Position the cursor arrow on the menu name **Draw** and *left-click* – the **Draw** pull-down menu appears. Position the pointer arrow on the word **Line** and *left-click*. A sub-menu appears. Position the pointer arrow on the word **Segments** and *left-click*. The pull-down menus disappear and the command line shows the prompt:

    LINE From point:

    The drawing of lines can now commence (Fig. 1.10);

2. *By selection from the on-screen menu*: position the pointer on the word DRAW in the AutoCAD on-screen menu. A sub-menu replaces the main on-screen menu. Select LINE from this sub-menu. The LINE sub-menu appears in its place. At the same time, the prompt:

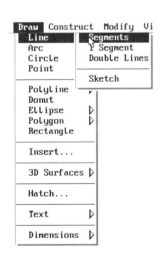

Fig. 1.10 The **Draw** pull-down menu

Fig. 1.11 The on-screen sub-menu for the **LINE** command

LINE From point:

appears at the command line. The drawing of lines can now commence (Fig. 1.11);

3. *By keying the command name from the keyboard*: in our example, type the word LINE at the keyboard and the command appears at the command line. *Left-click* and the prompt:

LINE From point:

appears at the command line. The drawing of lines can now commence.

4. *By keying abbreviations for command names from the keyboard*: a number of command name abbreviations are available in AutoCAD 12 depending upon the contents of the file *acad.pgp* usually held in the directory *acad\support*. The following is an extract from this file from the AutoCAD software loaded in my own computer:

A,	*ARC
B,	*BREAK
C,	*CIRCLE
CH,	*CHANGE
CP,	*COPY
DV,	*DVIEW
E,	*ERASE
H,	*HIDE
L,	*LINE
LA,	*LAYER
M,	*MOVE
MS,	*MSPACE
P,	*PAN
PS,	*PSPACE
PL,	*PLINE
R,	*REDRAW
S,	*STYLE
T,	*TEXT
VP,	*VPOINT
Z,	*ZOOM

There are other abbreviations in this file which will be explained later. From the list above it will be seen that to call the command LINE, all that is necessary is to key the letter L at the keyboard and the prompts:

**LINE
From point:**

appear at the command line and the drawing of lines can commence.

Notes

1. When keying commands from the keyboard either upper case (e.g. LINE or L) or lower case (line or l) can be used;
2. Throughout this book the method most frequently used will be to key abbreviations where possible at the command line;
3. When the command name is selected from a pull-down menu or keyed at the keyboard (whether in full or as an abbreviation), the sub-menu for the command appears in the on-screen menu.

The AutoCAD 12 on-screen menus

In the AutoCAD 12 on-screen menus, some of the command names will be followed by a colon (:), some by three fullstops (...), some with no colon or fullstops following the name. These three types of command names have the following meanings:

1. Those followed by a colon (:) e.g. **DIM:**. When these are selected by placing the cursor on the name followed by a *left-click*, a sub-menu associated with the command replaces the on-screen menu and the command appears at the command line, with or without prompts;
2. Those followed by three fullstops (...) e.g. **LAYER...**. When these are selected, a dialogue box appears on screen and the command line changes to include – in this example **DDLMODES**. **Note**: If **ddlmodes** is keyed into the command line, without first selecting **LAYER...** from the on-screen menu, the Layer dialogue box will also appear on screen;
3. Those which do not have a colon or fullstops after the name. When one such name is selected, a sub-menu will replace the existing on-screen menu and the command line will not change. **Note**: sub-menus resulting from such selections may contain commands followed by colons and may also include ddmode calls.

Note: when any command name is selected from an on-screen menu or sub-menu, that name highlights.

Some necessary settings

Setting limits

When working in two dimensions (2D), points in the AutoCAD drawing editor screen can be referred to in terms of two coordinate numbers in terms of x and y. Figures of the 2D coordinate position of the cursor in terms of x,y can be seen in the Status bar if **Coords** is on (the word **Coords** appearing in the Status bar). The point x,y = 0,0 is at the bottom left-hand corner of the drawing area of the editor. The x,y coordinates of the top right-hand corner of the drawing area depend upon figures set by the Limits command which can be selected from a pull-down menu – **Settings/Drawing Limits**, from an on-screen menu **SETTINGS/LIMITS:** or by typing **limits** at the keyboard:

 Command: limits (typed at the keyboard)
 ON/OFF/<Lower left corner><0.00,0.00>: *right-click*
 Upper right corner<12.00,9.00>: 420,297 (typed at
             keyboard) *right-click*
 Command:

This must be followed by:

 Command: zoom (typed at keyboard)
 All/Center/Dynamic/Extents/Left/Previous/Vmax/Window/
  <Scale(X/XP)>: a (all – typed at keyboard) *right-click*
 Command:

Why 420,297? Because the size of an A3 drawing sheet is 420 mm by 297 mm and for most of the exercises in this book we will be working in that size sheet.

Setting units

It will be seen from the above that there were two decimal places after the coordinate figures in the Limits prompts. The units the operator wishes to employ are best set by making use of the dialogue box **Units Control** selected from the **Settings pull-down menu**. This dialogue box is shown in Fig. 1.12. Note that Fig. 1.12 shows a feature of dialogue boxes not mentioned earlier – a *Pop-up list*, brought into the dialogue box by a *left-click* in the box at the top of the *Pop-up list*. When the list appears, a *left-click* on the required type of units to be displayed sets the units to the required decimal places. Personally I find two decimal places to be suitable when working in A3 sheet sizes. Note that, for the time being, only

Introduction 15

Fig. 1.12 The **Units Control** dialogue box

the decimal units have been set – as shown by the button against **Decimal** in the dialogue box being highlighted. To finally set the units to two decimal places, *left-click* on **OK** when the *Pop-up list* has closed.

Setting Grid and Snap

The two AutoCAD features Grid and Snap greatly facilitate ease and accuracy of drawing. When Grid is on, a series of dots will appear on screen at spacings depending upon how **Grid** has been set. When Snap is on, the cursor will snap onto snap-points at intervals which depend upon the setting of **Snap**. Both Grid and Snap can be set from the **Drawing Aids** dialogue box brought to screen by selecting **Drawing Aids** from the **Settings** pull-down menu – or from the **SETTINGS** on-screen menu – or by keying grid or snap at the keyboard and entering the required figures from the keyboard. Figure 1.13 shows the dialogue box with Snap set to x,y = 5,5 and with Grid set to x,y = 10,10. To set each figure, *left-click* in the appropriate box and enter the required figures from the keyboard. To ensure both settings are on check the **On** box by a *left-click* with the dialogue box arrow inside the check box in turn. If

Grid or Snap are to be off, a second *left-click* in the box will turn them off.

Coords

It will be seen from Fig. 1.13 that **Coords** can be turned on or off from within the Drawing Aids dialogue box. With **Coords** on, the coordinate numbers showing in the Status bar will be updated as the cursor is moved under the control of the mouse. **Coords** can also be turned on or off by typing the command **Coords** at the keyboard as follows:

> **Command:** coords (typed at keyboard)   *left-click*
> **Coords** <0>: 1 (typed at keyboard)   *left-click*
> **Command:**

Note: When set from the keyboard Coords has three settings as follows:

0: Coordinates appear in Status bar only at a point picked by the mouse;

1: Coordinates are constantly updated as the cursor is moved under the control of the mouse;

2: Coordinates show as *relative figures* as entities are constructed on screen. An entity is any single part of a drawing – a line, a circle, an arc, etc. More about relative coordinates in Chapter 2.

Ortho

It will be seen from Fig. 1.13 that another feature **Ortho** can be set on or off from within the **Drawing Aids** dialogue box. This feature

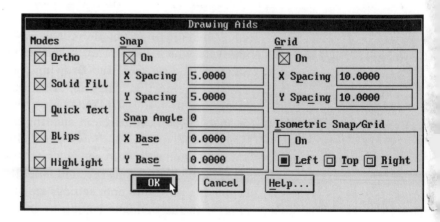

Fig. 1.13 The **Settings/Drawing Aids** dialogue box

can also be set by using key combinations or the function key F6. When set ON, all entities drawn in AutoCAD will be forced to follow strictly orthogonal routes – i.e. vertically up and down and horizontally left and right – all entities being drawn at right angles to the vertical or horizontal. This feature is of particular importance when constructing drawings which include orthographic projections or when entities must be constructed on orthogonal lines.

Function key calls

Shortcuts to the settings of some features on or off can be toggled by pressing functions keys:

F1: toggles between the Drawing editor and the text screen;
F6: toggles Coords on or off;
F7: toggles Grid on or off;
F8: toggles Ortho on or off;
F9: toggles Snap on or off.

In addition the following combinations of keys will perform the same functions:

Ctrl+O: toggles Ortho on and off;
Ctrl+G: toggles Grid on and off;
Ctrl+D: toggles Coords on and off;
Ctrl+B: toggles Snap on and off.

The use of Ctrl+C

One particularly important key combination is Ctrl and C. If a mistake is made when calling a command in AutoCAD, it can be cancelled by pressing these two keys together.

Revision

The following details were discussed in this chapter:

1. In this book we are dealing with AutoCAD Release 12 loaded into an IBM PC with at least a 386 operating system chip and at least 4 metabytes of memory and at least a VGA screen;
2. The system is working under MS-DOS 5.0;
3. AutoCAD 12 can be worked in Windows 3.1 if desired;
4. Check back on the advantages of using a CAD package for technical drawings (page 1);
5. Start up AutoCAD 12 with a batch file or other methods (page 3);

6. Opening a file from disk and opening a new file (page 6);
7. The AutoCAD drawing editor – Status bar, on-screen menu area, command line, pull-down menus;
8. Selection of command items – from on-screen menus; from pull-down menus; typing full command name; some abbreviations;
9. The acad.pgp file;
10. The meanings of a colon (:) and three fullstops (...) following a command name in on-screen menus;
11. The meaning of a triangular arrow after a command name in a pull-down menu and the meaning of three fullstops after a command name in a pull-down menu;
12. Cascading pull-down menus;
13. Dialogue boxes and their parts;
14. Calling dialogue boxes to screen – ddmodes – some examples are ddrmodes (drawing aids); ddunits (Unit control); ddmodify (brings on a modify dialogue box);
15. Ortho, Grid, Snap, Coords settings and toggles;
16. Setting Units, Limits, Grid and Snap values;
17. Function key calls to toggle screen, Grid, Snap, Ortho and Coords;
18. The AutoCAD 2D coordinate system;
19. Using Ctrl+C for command errors;
20. The meaning of the term *entity*;
21. The start-up drawing file *acad.dwg*.

A work disk

Most chapters in this book include exercises which are designed to provide the reader with a course of work, which if followed will allow him/her to attain a good degree of skill in operating AutoCAD for constructing technical drawings. If you are working with your own computer set-up, the files saved as a result of following this course can be stored on your own hard disk. If however you are working on a college computer or one shared with other operators, two problems arise:

1. If you store your work on the computer's hard disk, you will be taking up space on that disk which others may find a disadvantage to their own work;
2. If you use the standard set-up drawing file *acad.dwg* which appears on screen when you start-up AutoCAD, there is a possibility that the *acad.dwg* file may not be configured as a suitable drawing sheet for the exercises in this book.

Because of such difficulties you are advised to make up your own exercise drawing file, giving it a name which is personal to yourself and using this drawing file as the basic drawing sheet when working on exercises from this book. For the time being, until you become more proficient in using AutoCAD, it is suggested you make up a very simple file based on the features already discussed in this chapter. As your proficiency progresses, a file more suitable for more advanced work will be suggested later in this book. To set up the file proceed as follows:

1. Place a newly formatted disk in disk drive A:>\;
2. Start up AutoCAD. When the AutoCAD 12 files are fully loaded, select **Open** from the pull-down menu **File**;
3. Key in a suitable name in the edit file box of the dialogue box — e.g. if I were setting up this file I would name it a:\work;
4. When the Drawing editor re-appears set the following:
   (a) **Limits** to 420,297 (page 14);
   (b) **Grid** to 10 (page 15);
   (c) **Snap** to 5 (page 15);
   (d) From the **Units Control** dialogue box, set Grid and Snap ON;
   (e) Save the file to the filename, e.g. a:\work. The file will be saved to the disk in disk drive A:>\ with the filename *work.dwg*;
5. When you wish to construct an answer to an exercise, select **New** from the **File** pull-down menu and type the name of your file (a:\work) in the file edit box of the dialogue box. Do not forget the a:\. If you do, your work may eventually be saved to the hard disk of the computer you are using;
6. Construct your answer and save the file to disk drive a:\, with a filename such as a:\ex01.

Note: read the note below concerning the prototype file *acad.dwg*. You can easily check whether this prototype drawing loads a suitable drawing editor screen in which to construct answers to the exercises. Move the cursor with the mouse into the top right-hand corner of the screen to check whether the screen limits are indeed 420,297. Then press function key F7 to check whether a pattern of grid dots appear at intervals of 10 units vertically and horizontally. Finally press F9 to check whether Snap is set to a 5-unit interval. If all three checks prove satisfactory, there is no need to set up your own work file when working with AutoCAD Release 12. If using this book with earlier releases of AutoCAD, then you are advised to set up your own work file.

The prototype file *acad.dwg*

When AutoCAD starts up, a prototype drawing *acad.dwg* automatically loads. This drawing file is configured to produce a drawing editor screen suitable for the operators who usually use the computer into which AutoCAD is loaded. If this prototype *acad.dwg* produces a drawing editor suitable for working the exercises in this book, there is no need to set up your own work file drawing as described above.

CHAPTER 2

Constructing simple drawings

Introduction

Creating a new drawing

There are two main methods by which a new drawing can be started. These have been described earlier (page 7). The reader is advised to construct all drawing examples and exercises given here in order to build up skills in the construction of drawings in AutoCAD 12. In order to do so, it is probably best to create a new drawing for each example or exercise by selecting **New** from the **File** pull-down menu. The **Create New Drawing** dialogue box appears. Key in an appropriate drawing filename in the **New Drawing Name** box – e.g. the example given in Fig. 2.1 shows a file to be saved to a disk in disk drive **A:\>** with a filename ay01 (initials followed by a number). When the file is saved it will have a filename *ay01.dwg*. Do not include the extension .dwg when keying the name into the New Drawing Name box. The second file can be ay02 and so on. Use your own initials for this. You will then be able to readily see which files on your disks have been the result of working with this book.

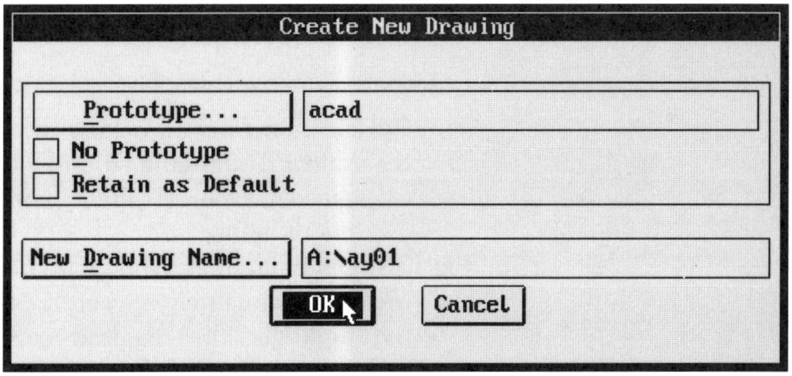

Fig. 2.1 The **Create New Drawing** dialogue box

Commands can be called by:

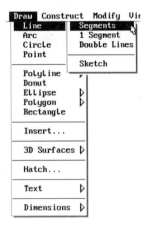

Fig. 2.2 The **Draw/Line** pull-down menu

1. Typing the full command name at the keyboard – the command name will appear at the command line as it is being typed, followed by the first of a series of prompts associated with the command;
2. Typing an abbreviation for the required command name if one exists (see page 12). The full command name appears at the command line, followed by the first of a series of prompts;
3. By selection from a pull-down menu:
    (i) position the cursor under mouse control in the Status bar;
    (ii) position the arrow which then appears over the pull-down menu name in the Status bar;
    (iii) *left-click*; the pull-down menu appears;
    (iv) *left-click* on the command name in the selected menu;

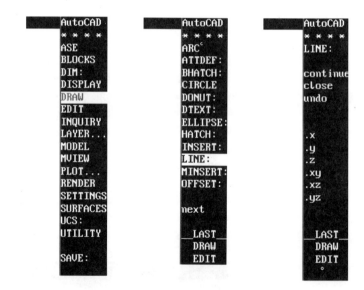

Fig. 2.3 The **DRAW/LINE** on-screen menus

4. By selection from an on-screen menu at the right-hand side of the AutoCAD drawing editor:
    (i) place the cursor under mouse control, over the required menu name;
    (ii) the name highlights; *left-click*;
    (iii) sub-menus may appear;
    (iv) highlight the required command name; the command name appears at the command line; *left-click*, a further

Constructing simple drawings 23

menu (a sub-menu) appears; prompts may be selected from the sub-menu if required.

As examples of the four methods given above, take as an example the command **LINE**. This is probably the most frequently used of all CAD commands. **LINE** can be called by either:

1. Typing LINE (or line) at the keyboard;
   or by:
2. Typing L (or l) at the keyboard;
   or by:
3. Selecting **Line** from the **Draw** pull-down menu (Fig. 2.2);
   or by:
4. Selecting **LINE:** from the **DRAW** on-screen menu (Fig. 2.3). Note that as with all on-screen menus, a second or third sub-menu may replace the AutoCAD main on-screen menu as selections are made. Note also that the selected command or prompt name highlights as the mouse is positioned over the on-screen menus.

Throughout this book we will be using the keyboard entry method of calling commands. If one is available from the *acad.pgp* file (page 12), the abbreviation will be shown followed by the full command name. If an abbreviation is not available, the full command name will be shown. It is up to the operator to decide which method best suits his or her method of working. After gaining experience in drawing with AutoCAD 12, most operators will find that some commands are best called by keying abbreviations, some by selection from pull-down menus and some from on-screen menus.

The current option prompt and choosing an option

All AutoCAD 12 commands appear at the command line with the options available with the command. For example, when the command **DLINE** is called the following appears at the command line:

Command: dl (dline)   *right-click (or Return)*
Break/Caps/Dragline/Offset/Snap/Undo/Width/<Start
                                                                point>:

This shows that there are options available – To Break; to add Caps at the line ends; to Drag the line; to Offset the line; to Undo the last line; and to alter the Width of the lines. From this set of prompts it is seen that <Start point> shows that the current option is to select the start point of the line. The statement in brackets, such as < >, indicates the current option. To select any one of the options all that is required is to key its initial – e.g. b for Break, followed by a *right-click*.

Entities

The word *entity* in AutoCAD 12 refers to a drawing feature such as a line, a circle, an arc, a word, etc.

Accuracy in the construction of drawings

Right from the start of commencing to learn how to use any CAD system, it is advisable to work using accurate constructional methods. For this reason, the majority of exercises and drawings in this book require the use of the AutoCAD 12 x,y (2D) drawing coordinate system. Later, when using 3D methods of construction, then the x,y,z coordinate system will be used.

Examples of x,y coordinate point positions in an AutoCAD 12 drawing editor with LIMITS set to 420,297 (A3 sheet size) are given in Fig. 2.4. Note that the bottom left-hand corner is x,y = 0,0. When a drawing constructed in this drawing editor is plotted full size (scale 1:1), each coordinate unit will plot as 1 mm.

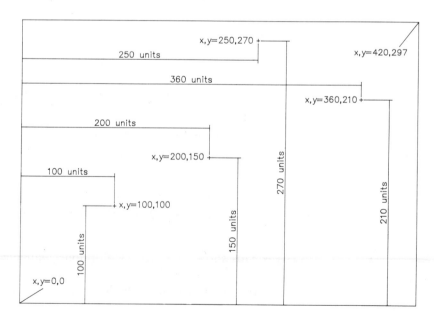

Fig. 2.4 Examples of x,y coordinate points in an A3 screen

Using the mouse buttons

Remember that throughout this book we will be employing a two-button mouse as the selection/pointing device. Pressing either the left or the right button will be described as either *left-click* or *right-click*.

A *right-click* usually has the same result as pressing the *Return* (or *Enter*) key of the keyboard, but occasionally pressing the *Return* key may have to be used instead of a *right-click*, or vice versa.

Points on the screen can be selected by placing the cursor, under mouse control, at the required position, followed by a *left-click* to confirm that position.

To select a menu item, place the cursor under mouse control over the menu item name and *leftclick*.

When only *Return* is shown against one of the items in the drawing sequences given here, it is essential that the *Return* key is pressed and not the right mouse button. However, pressing *Return* always produces the same result as a *right-click* of the mouse.

The command LINE

Figures 2.5, 2.6 and 2.7 show the three major methods by which accurate drawing of lines to given unit dimensions is achieved:

1. By keying the x,y figures of the *absolute coordinates* of points at line ends, at each **From point:** or **To point:** prompt of the **Line** command series of prompts;

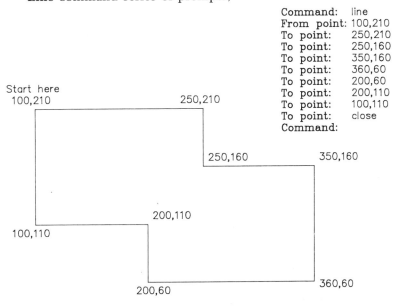

Fig. 2.5 Drawing with **LINE** – absolute coordinates

2. By keying x,y coordinates of points *relative* to the last point entered;
3. By keying angular x,y coordinates *relative* to the last point entered.

Absolute coordinates

Figure 2.5 shows a simple outline constructed by keying in the coordinate positions of the line ends. Follow the sequence below, entering the command abbreviation and the coordinate figures from the keyboard.

Command: l (line) entered at *keyboard* right-click (or Return)
From point: 100,210 *right-click*
To point: 250,210 *right-click*
To point: 250,160 *right-click*
To point: 350,160 *right-click*
To point: 360,60 *right-click*
To point: 200,60 *right-click*
To point: 200,110 *right-click*
To point: 100,110 *right-click*
To point: c (Close) *right-click*
Command:

The method of entering absolute coordinate points can be employed with most AutoCAD commands as will be seen later in this chapter.

Relative coordinates

Figure 2.6 shows another simple outline constructed by keying each coordinate point relative to the last coordinate entered. When

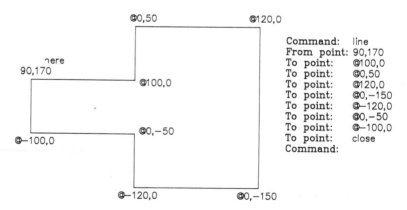

Fig. 2.6 Drawing with **LINE** – relative coordinates

Constructing simple drawings 27

using the relative coordinate method of entering points, the following rules must be observed:

1. The symbol @ must precede any absolute coordinate entered from the keyboard;
2. +ve x coordinates are to the right of the last coordinates entered;
3. −ve x coordinates are to the left of the last coordinates entered;
4. +ve y coordinates are above the last coordinates entered;
5. −ve y coordinates are below the last coordinates entered;
6. If the x coordinate value is 0, the line will be horizontal on screen;
7. If the y coordinate value is 0, the line will be vertical on screen.

To practise the use of absolute coordinates copy the drawing Fig. 2.6 by following the sequence:

Command: l (line) entered at *keyboard right-click* (or
                                                          Return)
From point: 90,170   *right-click* (or *Return*)
To point: @100,0   *right-click*
To point: @0,50   *right-click*
To point: @120,0   *right-click*
To point: @0,−150   *right-click*
To point: @−120,0   *right-click*
To point: @0,−50   *right-click*
To point: @−100,0   *right-click*
To point: c (Close)   *right-click*
Command:

Relative coordinates involving angles

Figure 2.7 shows a simple outline involving lines at angles to each other. When drawing such an outline using the relative coordinate

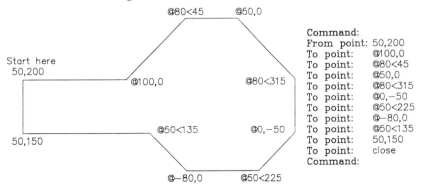

Fig. 2.7 Drawing with **LINE** – relative angular coordinates

method, the following rules must be observed:

1. The symbol @ must precede each x,y set of relative coordinate numbers;
2. The angle must be in degrees and be preceded by the symbol <;
3. Unless AutoCAD 12 has been configured otherwise, angles are measured from 0 degrees horizontally to the East (horizontally to the right), anti-clockwise through 90 degrees to the North (vertically upwards), through 180 degrees to the West (horizontally to the left) and through 270 degrees to the South (vertically downwards), back to 360 degrees (same position as 0 degrees).

Practise constructing an outline using angular relative coordinates by copying Fig. 2.7, following the sequence:

Command: l (line)   *right-click (or Return)*
From point: 50,200   *right-click*
To point: @100,0   *right-click*
To point: @80<45   *right-click*
To point: @50,0   *right-click*
To point: @80<315   *right-click*
To point: @0,−50   *right-click*
To point: @50<225   *right-click*
To point: @−80,0   *right-click*
To point: @50<135   *right-click*
To point: 50,150   *right-click*
To point: c (Close)   *right-click*
Command:

The command DLINE

Line from the **Draw** pull-down menu, allows the selection of **Double lines**. This can also be called as a command by entering dl (or dline) from the keyboard. Dline offers a variety of options shown by the following sequences (illustrated in Fig. 2.8):

Drawing 1

Command: dl (dline)   *right-click (or Return)*
Initializing... DLINE loaded
Break/Caps/Dragline/Offset/Snap/Undo/Width/<Start point>: w (width)   *right-click*
New DLINE width <0.50>: 5   *right-click*

Break/Caps/Dragline/Offset/Snap/Undo/Width/<Start point>: 35,280 *right-click*
Arc/Break/CAps/CLose/Dragline/Snap/Undo/Width/<next point>: 105,280 *right-click*
Arc/Break/CAps/CLose/Dragline/Snap/Undo/Width/<next point>: 105,210 *right-click*
Arc/Break/CAps/CLose/Dragline/Snap/Undo/Width/<next point>: 35,210 *right-click*
Arc/Break/CAps/CLose/Dragline/Snap/Undo/Width/<next point>: c (CLose) *right-click*
Command:

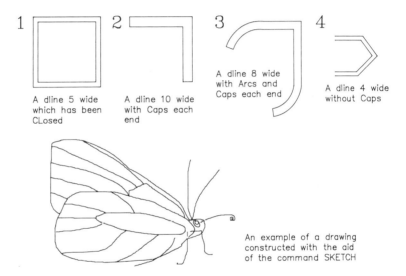

Fig. 2.8 Examples of drawings with **DLINE** and **SKETCH**

Drawing 2

Command: dl (dline) *right-click (or Return)*
Break/Caps/Dragline/Offset/Snap/Undo/Width/<Start point>: w (width) *right-click*
New DLINE width <5.0>: 10 *right-click*
Break/Caps/Dragline/Offset/Snap/Undo/Width/<Start point>: c (caps) *right-click*
Draw which end caps? Both/End/None/Start/<Auto>: b (both) *right-click*
Arc/Break/CAps/CLose/Dragline/Snap/Undo/Width/<Start point>: 135,275 *right-click*
Arc/Break/CAps/CLose/Dragline/Snap/Undo/Width/<next point>: 195,275 *right-click*

Arc/Break/CAps/CLose/Dragline/Snap/Undo/Width/<next
point>: 195,215   right-click
Arc/Break/CAps/CLose/Dragline/Snap/Undo/Width/<next
point>: Return (to finish command sequence)
Command:

Drawing 3

Command: dl (dline)   right-click (or Return)
Break/Caps/Dragline/Offset/Snap/Undo/Width/<Start
point>: w (width)   right-click
New DLINE width <10.0>: 8   right-click
Break/Caps/Dragline/Offset/Snap/Undo/Width/<Start
point>: c (caps)   right-click
Draw which end caps? Both/End/None/Start/<Auto>:
b (both)   right-click
Arc/Break/CAps/CLose/Dragline/Snap/Undo/Width/<Start
point>: 240,250   right-click
Arc/Break/CAps/CLose/Dragline/Snap/Undo/Width/<Start
point>: a (arc)   right-click
Arc/Break/CAps/CLose/Dragline/Snap/Undo/Width/
<second point>: 250,265   right-click
Arc/Break/CAps/CLose/Dragline/Snap/Undo/Width/
<Endpoint>: 270,270   right-click
Arc/Break/CAps/CLose/Dragline/Snap/Undo/Width/<next
point>: l (line)   right-click
Arc/Break/CAps/CLose/Dragline/Snap/Undo/Width/<next
point>: 310,270   right-click
Arc/Break/CAps/CLose/Dragline/Snap/Undo/Width/<next
point>: 310,210   right-click
Arc/Break/CAps/CLose/Dragline/Snap/Undo/Width/<next
point>: a (arc)   right-click
Arc/Break/CAps/CLose/Dragline/Snap/Undo/Width/
<second point>: 300,190
Arc/Break/CAps/CLose/Dragline/Snap/Undo/Width/
<Endpoint>: 285,185   Return end sequence
Command:

Drawing 4 of Fig. 2.8 is left as an example for readers to practise at this stage, without including the sequence of keyboard entries here.

Constructing simple drawings 31

The command SKETCH

Figure 2.8 includes an example of a drawing constructed by calling the command **Sketch** (see Fig. 2.2). The prompts showing at the command line when Sketch is called – either by selection from the **Draw/Line** pull-down menu or by entering sketch from the keyboard – are as follows:

Command: sketch   *right-click* (or *Return*)
Record increment <1.00>:   *right-click* (to accept this increment)
Sketch. Pen eXit Quit Record Erase Connect.

These options have the following meanings:

P – gives either **<Pen down>** or **<Pen up>** – toggling between the two as the key p is pressed;
eXit – type x and the command is cancelled;
Q – type q and the command ends;
R – type r and the outline sketch up to that moment changes from green to the drawing colour on screen;
E – type e and the last part of the sketch to be drawn is erased;
C – type c and a sketch line can be connected to an existing entity.

Note: when the command **Sketch** is used for freehand sketching:

1. The best results are probably obtained if a pen-type stylus is used as a selection/pointing device, although sketches can be produced with either a mouse or a puck;
2. With **Pen down**, movement of the selection/pointing device causes a sketch line to appear on screen. Typing P takes the pen up and a new sketch line can be started in a new position;
3. Drawing files containing constructions produced with this command tend to be large. This is because the command produces a series of straight lines, each of which requires memory to determine the positions of each end of the lines. The lengths of the lines are partly determined by the setting of the option **Record increment** within the sketch sequence.

The command ARC

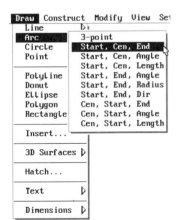

Fig. 2.9 The **Draw/Arc** pull-down menu

Arcs can be drawn in a variety of ways, as shown when the command **Arc** is selected from the **Draw** pull-down menu – Fig. 2.9. The same variety in abbreviated form will be seen in an on-screen sub-menu if **ARC:** is selected from the **DRAW** on-screen menu. As can be seen from these menus, arcs can be drawn by giving any three of the following:

>the arc start point;
>a second point on the arc;
>the end point of the arc;
>the arc's centre point;
>the angle made with the centre between the start point and the end point;
>the chord length of the arc (Length);
>the arc's radius.

A number of examples are given in Fig. 2.10. The sequences to follow to produce these arcs follow a pattern such as:

>**Command:** a (arc)   *right-click*
>**ARC Center/<Start point>:**   *pick* point on screen *left-click* or enter coordinates and *right-click*
>**Center/End/<Second point>:**   *pick* point on screen *left-click* or enter coordinates and *right-click*
>**End point:**   *pick* point on screen *left-click* or enter coordinates and *right-click*
>**Command:**

or:

>**Command:** a (arc) *keyboard*   *right-click*
>**ARC Center/<Start point>:** c (centre)   *right-click*

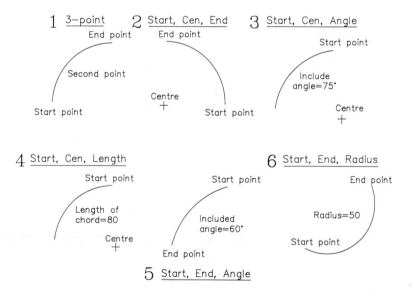

Fig. 2.10 Examples of drawings with **ARC**

Constructing simple drawings

Center: pick point on screen *left-click* or enter coordinates and *right-click*

Angle/Length of chord/<End point>: pick point on screen *left-click* (arc drags as mouse is moved)

Command:

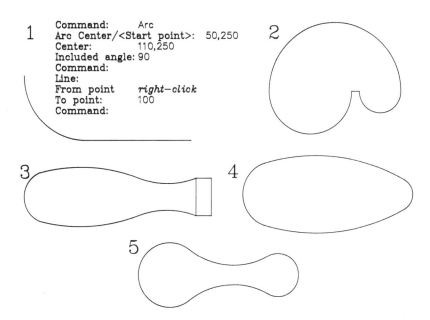

Fig. 2.11 Examples of outlines drawn with **ARC**

Drawing 1 of Fig. 2.11 shows a method of ensuring that a line is drawn which continues from the end of and is tangential to the arc. When **Line** is called and the **From point:** prompt is answered by *right-click* (or *Return*), the line resulting from entering coordinates at the **To point:** prompt is attached to the last entity drawn – in this example an arc. Drawings 2 to 5 are examples of simple outlines drawn with the command **Arc**.

The command CIRCLE

Circles can be drawn in a variety of ways as illustrated in Fig. 2.13. The command can be called from the pull-down menu **Draw** as shown in Fig. 2.12, from the on-screen menu **DRAW**, or by entering c (circle) from the keyboard. The examples in Fig. 2.13 were drawn by following the sequences:

Drawing 1

Command: c (circle) *right-click* (or *Return*)

3P/2P/TTR/<Center point>: 100,200   *right-click*
Diameter/<Radius>: 50   *right-click*
Command:

Drawings 2 and 3

Command: c (circle)   *right-click (or Return)*
3P/2P/TTR/<Center point>: 3p (three-point)   *right-click*
First point: 190,200   *right-click*
Second point: 240,240   *right-click*
Third point: 270,170   *right-click*
Command:

Drawing 3 was constructed in a similar fashion, but passing through only two points (prompt 2P).

Drawings 4 and 5

These two examples show the use of the **ttr** prompt. This prompt determines to which entities a circle is to be tangential.

Command: c (circle)   *right-click (or Return)*
3P/2P/TTR/<Center point>: 100,80   *right-click*
Diameter/<Radius>: 45   *right-click*
Command: *right-click*
3P/2P/TTR/<Center point>: 190,80   *right-click*
Diameter/<Radius>: 30   *right-click*
Command: *right-click*
3P/2P/TTR/<Center point>: ttr   *right-click (or Return)*
Enter tangent spec: *pick* point on circle R45 and *left-click*
Enter second tangent spec: *pick* point on circle R30 and
                                                              left-click
Radius 30.00:   *right-click* (to accept the radius of 30)
Command:

In the example drawing 4, the ttr prompt was used twice, with the radius for the lower circle being 20. In the example drawing 5, the ttr points were on the straight lines forming an angle to which the R40 circle is tangential.

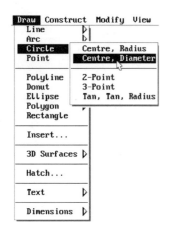

Fig. 2.12 The **Draw/Circle** pull-down menu

The command ELLIPSE

The pull-down menu **Draw** with the command **Ellipse** and its sub-menu are shown in Fig. 2.14. Figure 2.15 includes a number of

Constructing simple drawings 35

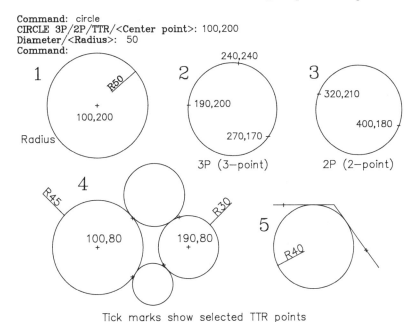

Fig. 2.13 Examples of drawings with the **CIRCLE** command

ellipses drawn with the use of the various options available with this command. The options of this command depend upon the fact that, in geometrical terms, an ellipse is centred on two axes – major and minor. The positions of the axes are shown in drawing 1 of Fig. 2.15. The sequences for two of the ellipses shown in Fig. 2.15 are as follows.

Drawing 2

Command: ellipse   *right-click* (or *Return*)
<Axis endpoint>/Center: 240,240   *right-click*
Axis endpoint 2: 340,240   *right-click*
<Other axis distance>/Rotation: 30 (or key the coordinates
                                  290,270)   *right-click*
Command:

Drawing 3

Command: ellipse   *right-click* (or *Return*)
<Axis endpoint>/Center: c (centre)   *right-click*
Center of ellipse: 290,150   *right-click*
Axis endpoint: 230,150   *right-click*
<Other axis distance>/Rotation: 25 (or key the coordinates
                                  290,175)   *right-click*
Command:

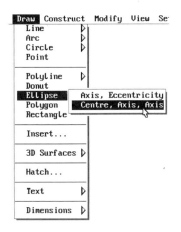

Fig. 2.14 The **Draw/Ellipse** pull-down menu

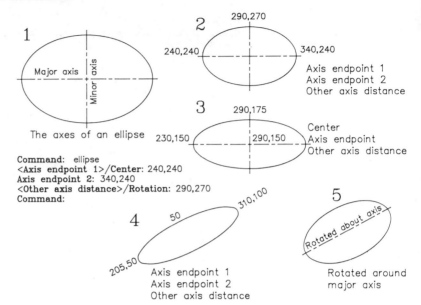

Fig. 2.15 Examples of ellipses drawn with the **ELLIPSE** command

```
Command: ellipse
<Axis endpoint 1>/Center: 240,240
Axis endpoint 2: 340,240
<Other axis distance>/Rotation: 290,270
Command:
```

Drawing 4 is an ellipse in which the axes have been chosen as being at an angle to horizontal — by keying coordinates not both on the same x-axis. Drawing 5 is an ellipse which has been drawn by using the r (Rotation about major axis) option. This option relies upon the fact that an ellipse is formed from a circle viewed from a frontal position and revolved around one of its diameters. As the revolution takes place the minor axis of the ellipse appears to change in length.

The command POLYGON

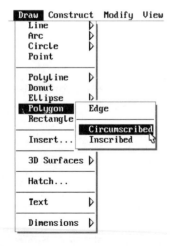

Fig. 2.16 The **Draw/Polygon** pull-down menu

Regular polygons — polygons with all sides the same length and all angles the same size — form an important part of many technical drawings. They are easily drawn in AutoCAD 12, with any number of sides from 3 upwards and either from stated edge lengths or inscribed/circumscribed within circles of stated radius/diameter. The **Draw/Polygon** pull-down menu is given in Fig. 2.16 and examples of polygons drawn with the command are shown in Fig. 2.17. The following sequences show the system of options available with the command.

Drawing 1

Command: polygon   *right-click* (or *Return*)

Constructing simple drawings 37

Number of sides <4>: 6
Edge/<center of polygon>: e (edge)   *right-click*
First endpoint of edge: 80,190   *right-click*
Second endpoint of edge: 130,190   *right-click*
Command:

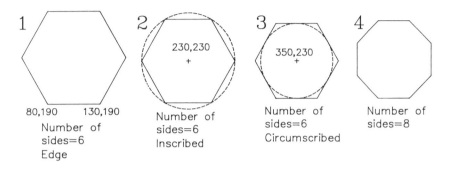

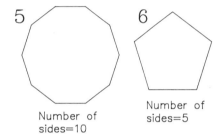

Fig. 2.17 Examples drawn with the command **POLYGON**

Fig. 2.18 The **PLINE** on-screen sub-menu

Drawing 2

Command: polygon   *right-click (or Return)*
Number of sides <4>: 6   *right-click*
Edge/<center of polygon>: 230,230   *right-click*
Inscribed in circle/Circumscribed about circle (I/C)<I>: i
                    (inscribed)   *right-click*
Radius of circle <50>: 50   *right-click*
Command:

Other examples are given in drawings 3 to 6.

The command PLINE (polyline)

Figure 2.18 shows the **PLINE:** on-screen sub-menu and Fig. 2.19 gives examples of plines drawn with this command. Because plines will be dealt with later in this book, only the details of drawing a simple straight line pline of width 0.7 units is given in the sequence example below.

38 AutoCAD Release 12 for students

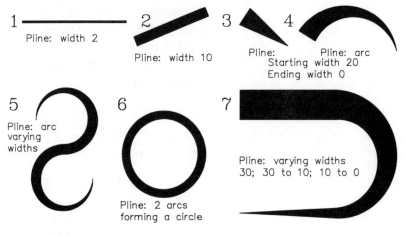

Fig. 2.19 Examples drawn with the command **PLINE**

```
Command: pline
From point: 35,280
Current width is 2.00
Arc/Close/Halfwidth/Length/Undo/Width/<Endpoint>: 145,280
Command:
```

In the construction of industrial drawings such as engineering and building drawings polylines are important because, with the aid of the command, lines of different thicknesses can be plotted from drawings in AutoCAD. Examples of straight line pline drawings will be given in a few exercises in this chapter. The pline options are shown in the following sequence:

Command: pl (pline)   *right-click (or Return)*
From point: 100,50   *right-click (or Return)*
Current width is 0.00
Arc/Close/Halfwidth/Length/Undo/Width/<Endpoint of
                              line>: w (width)   *right-click*
Starting width <0.00>: 0.7   *right-click*
Ending width <0.700>:   *right-click (or Return to accept)*
Arc/Close/Halfwidth/Length/Undo/Width/<Endpoint of
                              line>: 150,200   *right-click*
Arc/Close/Halfwidth/Length/Undo/Width/<Endpoint of
                              line>: 150,200   *right-click*
Arc/Close/Halfwidth/Length/Undo/Width/<Endpoint of
                              line>:   *right-click*
Command:

Note: figure 2.20 demonstrates what happens when a polyline is drawn with the variable **FILL** set to off. When drawing polylines, if they are to show on screen and plot as solid lines the variable **FILL** must be on.

Constructing simple drawings

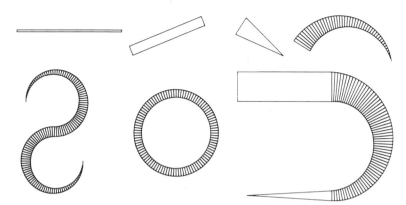

Fig. 2.20 The plines of Fig. 2.19 with **FILL** off

Command: fill   *right-click*
FILL ON/OFF<ON>:   *right-click* (to accept)
Command:

Note: If an arc is to be drawn it can be constructed to the same parameters as when the command **ARC** is called. This is shown when a (Arc) is the response to prompts at the **PLINE** command line:

Command: pl (pline)   *right-click*
From point: *pick*
Arc/Close/Halfwidth/Length/Undo/Width/<Endpoint of line>: a (arc)   *right-click*
Angle/Center/CLose/Direction/Halfwidth/Line/Radius/ Second pt/Undo/Width/<Endpoint of arc>:

Key the capitals of the options as required to draw the required arc.

Adding text to drawings

In order to add text – titles, notes etc. to drawings – it is necessary to first set the text **style**. The standard AutoCAD 12 package contains a large number of text fonts, any of which can be set as the current font by calling the command **Style**. If, for example, one wishes to set the font **Romanc**:

Command: style   *right-click*
STYLE Text style name (or ?)<Simplex>: romanc   *right-click*

This brings the **Select Font Style** dialogue box on screen (Fig. 2.21). The required font can be selected from the dialogue box by placing

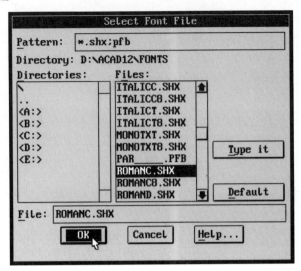

Fig. 2.21 The **Select Font Style** dialogue box

the box arrow over the name of the required font followed by a *left-click*. The selected name appears in the **File:** box. *Left-click* on **OK** and the first of a series of six prompts appears at the command line. The height, width etc. of the font can be from these prompts:

> **Height <0.00>:** 6   *right-click*
> **Width factor <1>:**   *right-click* to accept the width factor 1
> **Obliquing angle <0>:**   *right-click* to accept the angle 0 degrees
> **Backwards ? <N>:**   *right-click* to accept the obliquing angle of 0 degrees
> **Upside-down ? <N>:**   *right-click* to accept No
> **Vertical ? <N>:**   *right-click* to accept No
> **ROMANC is now the current style**
> **Command:**

A number of the fonts available from AutoCAD 12 are given in Fig. 2.22. These have been set from the **Style** command prompts to a variety of heights, widths, obliquing angles, etc.

Notes

1. Not all the fonts available in AutoCAD 12 are given in Fig. 2.22;
2. A number of Postscript fonts are available in AutoCAD 12. Some of these are shown in Fig. 2.22.

By repeated use of the **Style** command, a number of fonts of varying sizes can be set up in any one drawing. Once a number of

Constructing simple drawings   41

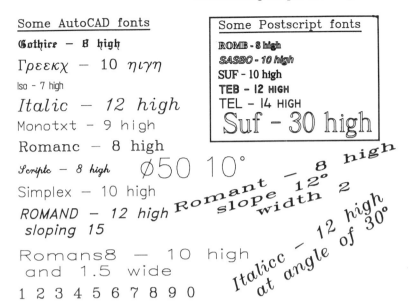

Fig. 2.22 A selection of fonts from AutoCAD 12

fonts have been set, in order to select the font the operator wishes to use, the **Entity Creation Modes** dialogue box can be selected from the **Settings** pull-down menu. Then point at and *left-click* on the **Text Style...** box. The **Select Font Style** dialogue box appears. If, for example, the **ROMANC** font name is selected and **Show All ...** pointed at, followed by a *left-click*, the appearance of the selected font can be seen in another dialogue box **Text Style – ROMANC Symbol Set**. This is shown in Fig. 2.23.

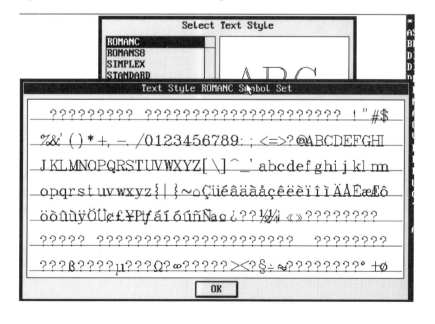

Fig. 2.23 The **Text Style Symbol Set** dialogue box

An existing style can also be set by use of the **Style** command and following the prompts which appear.

Adding text

Text can be added to a drawing by calling the **TEXT** command. This gives rise to a number of prompts allowing the operator to set the position and slope angle of the text to be placed. A number of examples of settings, using the font style **SIMPLEX**, of height 8 units, are given in Fig. 2.24. When the command **TEXT** is called, the following appears at the command line:

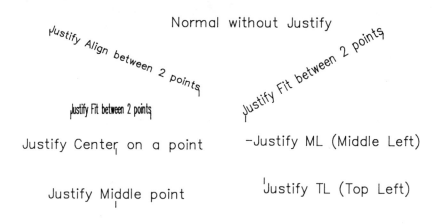

Fig. 2.24 Examples of methods of placing text

Command:
TEXT Justify/Style/<Start point>: *pick* start point or key a coordinate
Rotation angle <0>: *right-click* to accept rotation angle of 0 degrees
Text: type required text   *right-click*
Command:

and the text appears commencing at the selected **Start point**.

If the text is to be placed in positions other than from a selected start point or within given points, use the prompt **Justify** as follows.

Command: text   *right-click*
TEXT Justify/Style/<Start point>: j (justify)   *right-click*
Align/Fit/Center/Middle/Right/TL/TC/TR/ML/MC/MR/BL/BC/BR:

Constructing simple drawings

Followed by prompts associated with the selected option, for example for the option a (align):

First text line point: *pick* or key a coordinate
Second text line point: *pick* or key a coordinate
Text: type the required text   *right-click*
Command:

and the text appears on screen to the required justification.

A work sheet

Figure 2.25 shows a simple work sheet which could be drawn each time an example or exercise from the book is attempted. This uses the *work.dwg* file described on page 19, with added borders and titles. The titles have been added using the **Simplex** font set at 8 units high. The borders are as large as the work.dwg sheet allows; the title block is 20 units high.

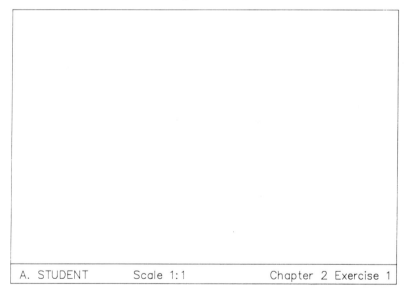

Fig. 2.25 A suggested work sheet for exercises

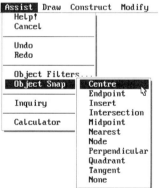

Fig. 2.26 The **Assist/Object Snaps** pull-down menu

Object snaps

The setting of **Object snaps** (osnaps) allows the exact positioning of entities onto other entities in a construction. When osnaps are in action, a **pick-box** will be seen at the junction of the cursor hairlines. The start or other points for positioning of an entity can be precisely placed by positioning the **pick-box** at the required point on an entity already on the screen. The end of the new entity then snaps onto the selected point. The variety of types of osnaps

Fig. 2.27 **Osnaps** from on-screen menus

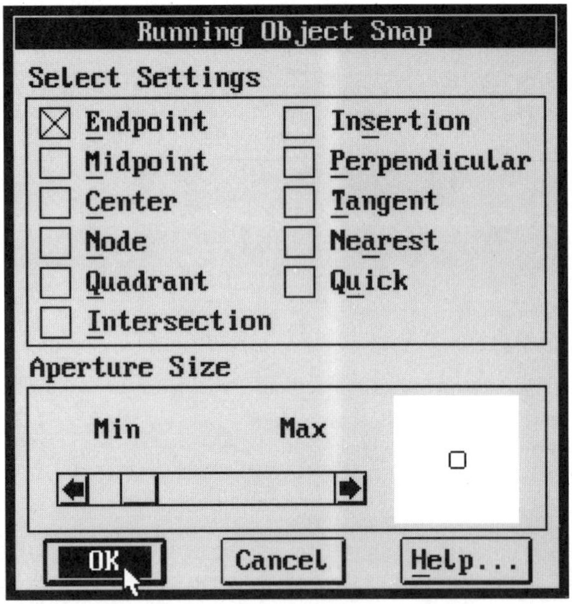

can be seen by selecting **Object Snap ...** from the **Assist** pull-down menu (Fig. 2.26). As can be seen in this illustration, entities can be made to snap onto a number of features already on screen. Of those shown in Fig. 2.26, Insert will be understood when Inserts are dealt with later (page 135).

Osnaps can be set for just each snap point operation by selection either from the **Object Snap** pull-down menu or by selecting the four stars **** from the AutoCAD main on-screen menu. When **** is selected, the osnap sub-menu appears in its place (Fig. 2.27). Selecting, say, **ENDpoint** from this sub-menu allows the entity being drawn to snap onto the end point of a selected entity already on screen.

Fig. 2.28 The **Running Object Snap** dialogue box

Osnaps can be set for each and every snap point operation by selecting **Object Snap ...** from the **Settings** pull-down menu. The **Running Object Snap** dialogue box appears (Fig. 2.28). Any one, several or all the osnaps can be set by a *left-click* within the box(es) against the osnap name. The size of the pick-box can also be changed within the dialogue box by movement of the slider bar towards either **Min** (to decrease its size), or **Max** (to increase its size).

Constructing simple drawings   45

Notes

1. If **Snap** is on and the construction being developed depends upon sizes which are always multiples of the snap setting, precise positioning of the ends of entities in relation to each other can usually be achieved without resorting to osnaps;
2. Trying to accurately select points at the ends, centre of, or other points of entities on screen is impossible without the aid of either snap or osnap. When positioning entities without the aid of these two facilities, it may appear to the operator that accurate positioning has been achieved without them, but this is always a fallacy.

Notes on lines of text and the command DTEXT

If the following sequence is followed when adding text to a drawing, the second text typed at the keyboard is placed on screen immediately below the first:

> **Command:** text   *right-click*
> **TEXT Justify/Style/<Start point>:**   *pick* start point or key a coordinate
> **Rotation angle <0>:**   *right-click* to accept rotation angle of 0 degrees
> **Text:** type first line of text   *right-click*
> **Command:**
> **Rotation angle <0>:**   *right-click* to accept rotation angle of 0 degrees
> **Text:** type second line of text   *right-click*
> **Command:**

Further lines of text can be added in lines one below the other by continuing this sequence.

If, instead of using the command **TEXT**, the command **DTEXT** is called, the text appears on screen as it is being keyed at the keyboard. Some operators prefer using dtext, because the text can be seen on screen as it is typed, allowing features such as correct choice of font and a suitable height to be seen at once before a long line of text is entered.

Exercises

The following simple exercises are designed to allow the reader to practice drawing with the aid of the commands and facilities

Exercise 1

Working to the dimensions given in Fig. 2.29, copy the drawings 1 to 4, using the LINE command, with snap set to 5 and set on. Do not attempt to add the dimensions.

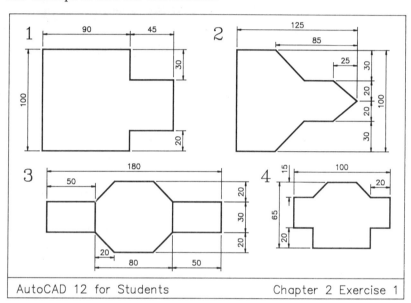

Fig. 2.29 Exercise 1

Exercise 2

With the aid of the three commands LINE, CIRCLE and ARC, construct drawings 5 to 8 of Fig. 2.30. Make sure snap is on at a setting of 5. Do not attempt adding the dimensions.

Exercise 3

Using the command DLINE, copy the four drawings 9 to 12 of Fig. 2.31. Snap should be set to 5 and on. Do not include any of the dimensions.

Constructing simple drawings   47

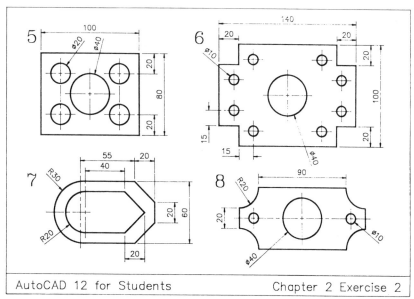

Fig. 2.30 Exercise 2

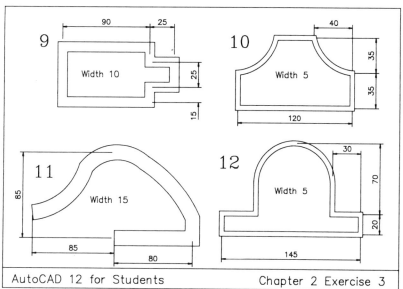

Fig. 2.31 Exercise 3

Exercise 4

With the commands CIRCLE and LINE copy the given two drawings of Fig. 2.23. The circles are either tangential to other circles (drawing 13) or tangential to straight lines (drawing 14). Do not include any dimensions.

Fig. 2.32 Exercise 4

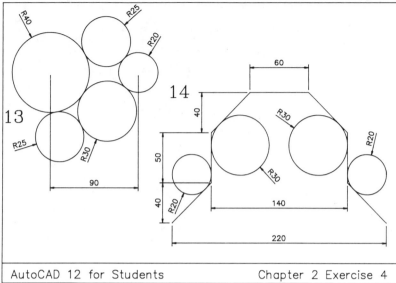

Fig. 2.33 Exercise 5

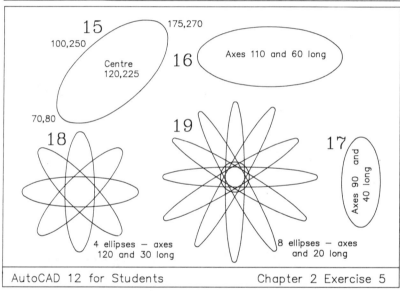

Exercise 5

Using the command ELLIPSE construct the five outlines – drawings 15 to 19 – of Fig. 2.33. Do not include the dimensions.

Exercise 6

Construct the five plines of Fig. 2.34, working to the information given with the drawings. The dimensions are for guidance only and should not be included in your answer.

Constructing simple drawings 49

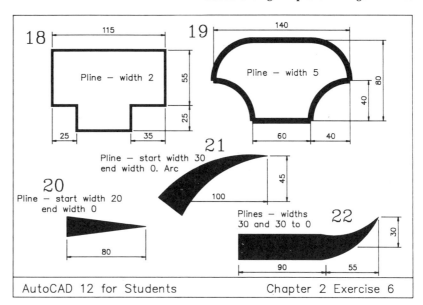

Fig. 2.34 Exercise 6

Exercise 7

Figure 2.35 is a simple design involving text, PLINEs and DLINEs. Copy the given drawing, without including the information concerning the text and lines.

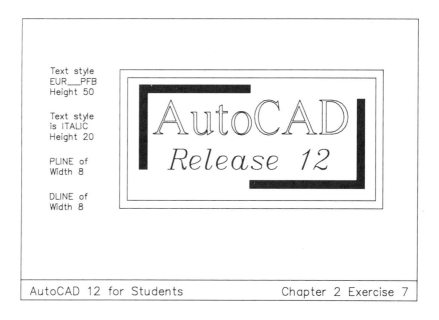

Fig. 2.35 Exercise 7

Exercise 8

With the command POLYGON construct the drawings of Fig. 2.36. Do not include any of the notes.

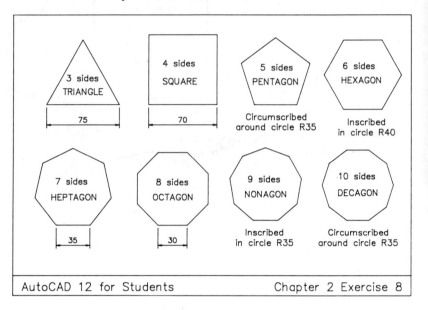

Fig. 2.36 Exercise 8

CHAPTER 3

Modifying drawings

Introduction

The commands in the **Modify** pull-down menu can be seen in Fig. 3.6 (page 56). With the **Zoom** command from the **View** menu (Fig. 3.1), these have a common set of prompts associated with the selection of entities to be modified. These prompts do not necessarily always appear when most of the commands are called, but are available if required.

Take as an example the command **MOVE**. If m (abbreviation for Move) is entered at the keyboard; **Move** selected from the **Modify** pull-down menu; or **MOVE:** selected from the on-screen menu, the following appears at the command line:

>**Command:** m (move)   *right-click*
>**MOVE**
>**Select objects:**

A small box – the *object selection target* box – replaces the cursor cross-hairs. If this is moved under mouse control onto an entity and is followed by a *left-click* (*picked*) the selected entity highlights and the following prompt appears:

>**Select objects:**   *right-click*
>**Base point of displacement:**   *pick* or key coordinates
>**Second point of displacement:**   *pick* or key coordinates
>**Command:**

and the selected entity is moved to the selected second point of displacement.

At the **Select objects:** prompt – common to all **Modify** and **Zoom** commands responses, other than to select a single entity, are possible: e.g. w (abbreviation for Window) – as in the following example:

>**Command:** m (move)   *right-click*

MOVE
Select objects: w (window)
First corner: *pick* or key coordinates
Other corner: *pick* or key coordinates **5 found**
Select objects: *right-click*
Base point of displacement: *pick* or key coordinates
Second point of displacement: *pick* or key coordinates
Command:

and the entities within the window box are moved to the second point of displacement.

Other possible prompts, which do not always appear at the command line, unless an incorrect response is made – e.g. if o is entered – are from the set:

**Window/Last/Crossing/BOX/ALL/Fence/WPolygon/
CPolygon/Add/Remove/Multiple/Previous/Undo/Auto/
Single**

The responses to each of these prompts is given by entering the initial letter of the prompt, e.g. **w** for Window, **l** for Last – or the capital letters contained in the prompt – e.g. **box** and not b; **all** and not a; **cp** and not c. The prompts have the following meanings:

Window: selects all the entities completely within the area of a rectangle formed from **First corner:** and **Other corner:**
Last: select the last entity to be drawn;
Crossing: selects all entities crossed by the lines of a rectangle formed from **First corner:** and **Other corner:**
BOX: if the **Other point:** of the window rectangle is to the right of the **First corner:** BOX has the same effect as Window. If the **Other point:** is to the left of the **First point:** BOX has the same effect as Crossing;
ALL: selects all entities in the drawing;
Fence: allows a set of selection lines to be drawn as a continuous "fence". All entities within or crossed by the fence become selected;
WPolygon: allows a polygon to be drawn to enclose the entities to be selected. This form of selection polygon cannot contain lines which cross within the polygon;
CPolygon: similar to both the WPolygon and the Crossing prompts. The lines of the polygon can cross and all entities both within the polygon and which the polygon's lines cross are selected;
Add: entities can be added to the set already selected;
Remove: entities can be removed from the entities already selected;

Modifying drawings 53

Multiple: e.g. allows multiple copies to be made;
Previous: to re-select the last selected entity or set of entities;
Undo: undoes the last selection;
AUto: will automatically form the First point of a crossing window
   if a selection is made at a point where there is no entity;
SIngle: to select a single entity.

Note: we will only be referring to the following prompt responses in this book: **W**indow, **C**rossing, **M**ultiple, **P**revious. This is because they are most likely to be used when employing the Modify and Zoom command systems. In fact, when these AutoCAD 12 commands are called, most of the set of prompts shown above do not appear, as will be seen in the examples in this chapter.

The command ZOOM

ZOOM is one of the most frequently used commands in any CAD system. This frequency lies in the fact that the smallest area of a drawing on screen can be viewed for additions or modification by "zooming" in or out of the current drawing or part of a drawing on screen. As with other commands in AutoCAD 12 it can be called by selection from a pull-down menu (**View**) – Fig. 3.1; from an on-screen menu; keyed in full; or keyed as the abbreviation z at the command line. When called, the following appears at the command line:

Command: z (zoom)   *right-click*
**All/Center/Dynamic/Extents/Left/Previous/Vmax/Window/
Scale(X/XP):**

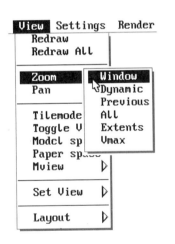

Fig. 3.1 The **View** pull-down menu

Examples of zooming to windows to scale by accepting the current prompt (**X/XP**) and to the drawing extents are given below. The reader is advised to experiment with other prompts in the zoom command as the opportunity arises.

Figure 3.2 is an orthographic two-view drawing within an AutoCAD 12 drawing editor screen. Figure 3.3 shows the effects of responses to the window, scale and the extents prompts.

Drawing 1

The zoom window formed when the command line shows:

Command: z (zoom)   *right-click*
**All/Center/Dynamic/Extents/Left/Previous/Vmax/Window/
Scale(X/XP):** w (window) entered at keyboard   *right-click*

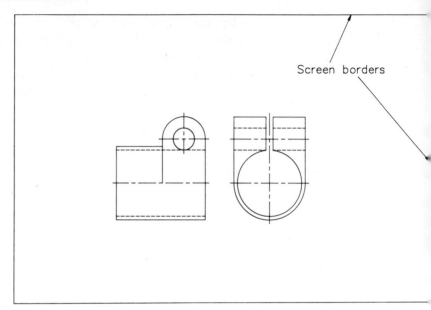

Fig. 3.2 A two-view drawing of a pipe clip

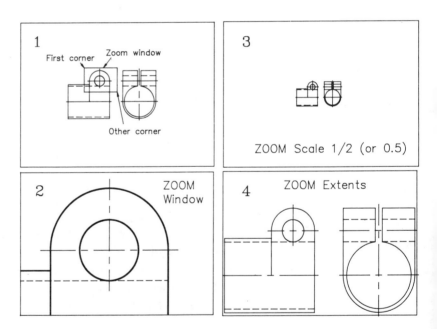

Fig. 3.3 A **ZOOM** Window, a **ZOOM** Extents and a **ZOOM** Scale

First corner: *pick* point on screen or enter coordinates
Other corner: *pick* point on screen or enter coordinates
Command:

and the screen changes as in **Drawing 2**.

Modifying drawings

Drawing 3

The screen resulting from the following:

> **Command:** z (zoom) *right-click*
> **All/Center/Dynamic/Extents/Left/Previous/Vmax/Window/**
>   **Scale(X/XP):** 1/2 (or 0.5) entered at keyboard *right-click*
> **Command:**

Drawing 4

The zoom window formed when the command line

> **Command:** z (zoom) *right-click*
> **All/Center/Dynamic/Extents/Left/Previous/Vmax/Window/**
>   **Scale(X/XP):** e (extents) entered at keyboard *right-click*
> **Command:**

The power of ZOOM

Figures 3.4 and 3.5 illustrate how powerful the command ZOOM can be. Even these two illustrations, however, do not show the largest scale of zoom possible. If necessary AutoCAD 12 can zoom to a much larger scale than the 2000:1 shown by these two drawings.

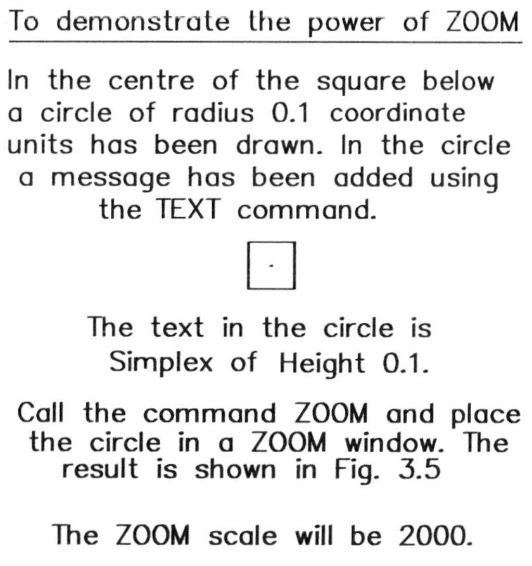

```
To demonstrate the power of ZOOM

In the centre of the square below
a circle of radius 0.1 coordinate
units has been drawn. In the circle
a message has been added using
the TEXT command.

         ┌───┐
         │ · │
         └───┘

The text in the circle is
Simplex of Height 0.1.

Call the command ZOOM and place
the circle in a ZOOM window. The
result is shown in Fig. 3.5

The ZOOM scale will be 2000.
```

Fig. 3.4 The power of **ZOOM** (Drawing 1)

The Modify menu commands

Figure 3.6 shows the commands in the **Modify** pull-down menu. As with other commands in AutoCAD 12 these can all be called either

This text has been added here using AutoCAD 386 Release 12.

The text is Simplex of height 0.1.

Fig. 3.5 The power of **ZOOM** (Drawing 2)

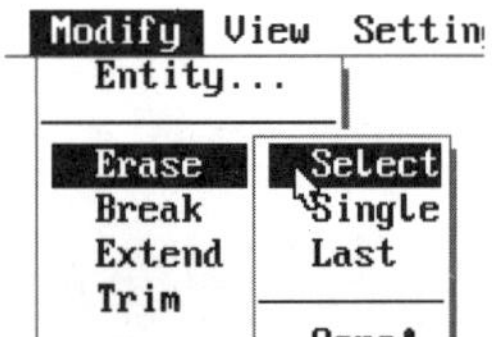

Fig. 3.6 The **Modify** pull-down menu

by entering the command (or its abbreviation) at the keyboard, by selection from the pull-down menu or from an on-screen menu. The first of the commands in the **Modify** on-screen menu is **Entities** .... To explain what an entity is refer to Fig. 3.7. One particular entity seen in Fig. 3.7 and not shown earlier is that formed by using the **POINT** command. The **Point Style** dialogue box (Fig. 3.8) will come on screen by selection from the **Settings** pull-down menu. To select the desired Point Style move the dialogue box arrow on to the picture of the required point style and *left-click*. When **POINT** is called and a *left-click* made at a point on screen the chosen Point Style will display at that point.

Modifying entities

If necessary, entities on screen can be modified by selecting **Entity** ... from the **Modify** pull-down menu. A prompt at the command line requests that the entity to be modified be selected. The **Modify** dialogue for the particular entity type appears on screen. Changes can be made by selection within the boxes in the dialogue box or by keying new coordinates, angles, radii, etc., in the appropriate boxes in the dialogue box. The dialogue box for arc

Modifying drawings 57

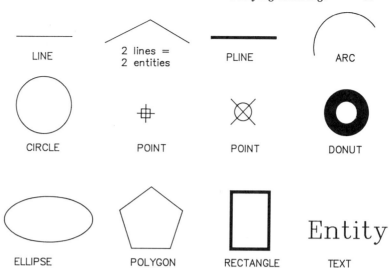

Fig. 3.7 Examples of entities

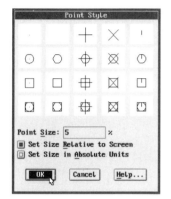

Fig. 3.8 The **Point Style** dialogue box

Fig. 3.9 The **Modify Arc** pull-down menu

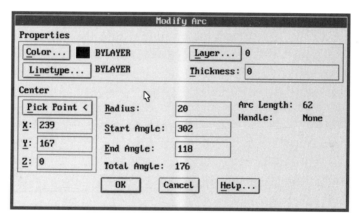

is shown in Fig. 3.9. Other Modify dialogue boxes are available for other entities (objects) such as lines, plines, circles, etc.

The command ERASE

Erasure of unwanted objects from the screen is achieved by calling the command **ERASE**. If a single entity or several entities are to be erased (Fig. 3.10) the command line shows:

>**Command:** e (erase)   *right-click*
>**ERASE**
>**Select objects:**   *pick* an entity to be erased **1 found**

Select objects: pick an entity to be erased **1 found**
Select objects: right-click
Command:

and the two entities disappear from the screen.

If a group of entities are to be erased, either a window or a crossing window can be used for the purpose. Figure 3.11 illustrates these two methods. The command line will show for each of these:

Drawings 1 and 2

Command: e (erase)   right-click
ERASE
Select objects: w (window)   right-click
First corner: pick first window corner on screen
Other corner: pick other window corner on screen
Select objects: right-click
Command:

and the entities within the window disappear from the screen.

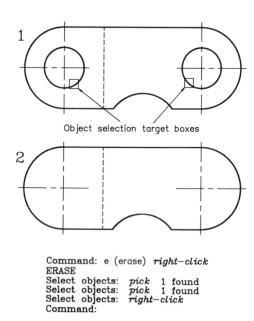

```
Command: e (erase) right-click
ERASE
Select objects:  pick  1 found
Select objects:  pick  1 found
Select objects:  right-click
Command:
```

Fig. 3.10 An example of erasing entities

Drawings 3 and 4

Command: e (erase)   right-click
ERASE

Modifying drawings

>**Select objects:** a (crossing window)   *right-click*
>**First corner:**   *pick*   **Other corner:**   *pick*
>**Select objects:**   *right-click*
>**Command:**

and the entities crossed by the outline of the crossing window will be erased.

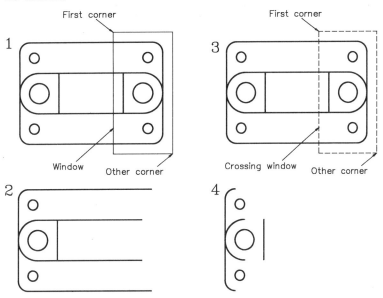

Fig. 3.11 Examples of erasing by a window and a crossing window

The command MOVE

The **MOVE** command allows the moving of single entities (objects) or groups of entities. In the examples given here, groups of entities are shown as being moved, but the same procedures are followed when moving single entities. For example, it is often necessary to move text to a better position in relation to the drawing as a whole. A line of text is treated as a single entity.

When **MOVE** is called from the **Modify** pull-down or on-screen menu or keyed at the command line, the following appears at the command line:

>**Command:** m (move)   *right-click*
>**Select objects:** w (window)
>**First corner:**   *pick* or key coordinates
>**Other corner:**   *pick* or key coordinates **5 found**
>**Select objects:**   *right-click*
>**<Base point of displacement>:**   *pick* or key coordinates
>**Second point of displacement:**   *pick* or key coordinates
>**Command:**

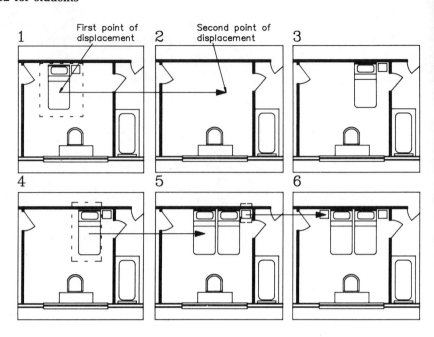

Fig. 3.12 Examples of **MOVE** and **COPY**

Fig. 3.13 Copy in the **Construct** pull-down menu

and the group of entities is moved to the second point of displacement.

The drawings 1, 2 and 3 of Fig. 3.12 show the stages in the moving of part of a drawing from one position to a second. Each drawing in Fig. 3.12 is part of the plan of a bungalow showing the position of furniture and fitments – only a bedroom and part of the attached bathroom are shown. The bed and bedside cabinet have been placed in an awkward position – too near the bedroom door – and should be moved in the plan drawing. The sequence involved in moving the two items of furniture is as follows:

Drawing 1. *Pick* the two corner points of a window around the objects to be moved. *Pick* a suitable point of displacement. This can be either any point in the window or any point on the screen. It may at times be easier to select a point outside the window;

Drawing 2. Move the selection device. As it is moved a ghosted copy of the windowed objects moves on the screen in response to the selection device movements. The objects are being **DRAG**ged across the screen;

Drawing 3. *Right-click* and the **MOVE** has been completed.

The command COPY

The command **COPY** will not be found in the **Modify** set of commands. To call **COPY** from a pull-down menu, select the

Construct menu – Fig. 3.13. The reason for including this command at this stage is because its action is similar to the **MOVE** command, as can be seen from drawings 4, 5 and 6 of Fig. 3.12. The major differences between the two commands is that with the **COPY** command, the object being copied remains in its position on screen and several copies of the same objects can be made with the aid of the **Multiple** option. The example given in Fig. 3.13 uses the same bed and bedside cabinet example. After these have been moved (shown by drawings 1, 2 and 3), a second bed and cabinet are copied as follows:

Drawing 4. Window the bed that has been moved. *Pick* a suitable base point of displacement (either in or outside the window);

Drawing 5. *Pick* a second point of displacement and the bed plan is copied. Window the bedside cabinet plan and *pick* a first point of displacement for the cabinet plan;

Drawing 6. *Pick* a second point of displacement and a copy of the cabinet plan is placed in position.

The sequence of prompts appearing at the command line for **COPY** follows the pattern:

> **Command:** cp (copy)   *right-click*
> **COPY**
> **Select objects:** w (window)
> **First corner:**   *pick* or key coordinates
> **Other corner:**   *pick* or key coordinates **5 found**
> **Select objects:**   *right-click*
> **<Base point of displacement>/Multiple:**   *pick* or key coordinates
> **Second point of displacement:**   *pick* or key coordinates
> **Command:**

A second example involving multiple copies is given in Fig. 3.14, in which a drawing of a slotted hole is copied three times. The sequence of prompts and responses to produce the three copies were:

> **Command:** cp (copy)   *right-click*
> **COPY**
> **Select objects:** w (window)   *right-click*
> **First corner:**   *pick* or key coordinates
> **Other corner:**   *pick* or key coordinates **8 found**
> **Select objects:**   *right-click*
> **<Base point of displacement>/Multiple:** m (multiple)   *right-click*

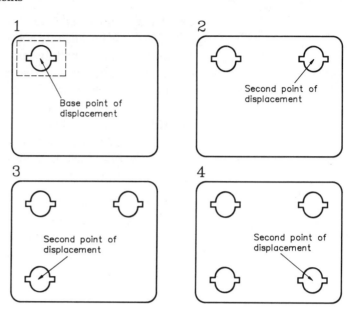

Fig. 3.14 An example of Multiple copies

      Base point: *pick* or key coordinates
      Second point of displacement: *pick* or key coordinates
      Second point of displacement: *pick* or key coordinates
      Second point of displacement: *pick* or key coordinates
      Second point of displacement: *right-click*
      Command:

The command ROTATE

With the aid of **ROTATE**, a drawing or any part of a drawing can be rotated through any angle. When this command is in action it must be remembered that the common configuration for AutoCAD produces an anti-clockwise angular movement. Note however that rotation can be effected in a clockwise direction if the setting in the **Direction Control** dialogue box (**Settings/Units Control**) is changed to that effect.

    Seven of the drawings of Fig. 3.15 show the results of rotating an arrow, drawn with lines, through 45-degree intervals. The command line sequence for each of the rotations is:

      Command: rotate   *right-click*
      Select objects: a (window)   *right-click*
      First corner: *pick* or key coordinates
      Other corner: *pick* or key coordinates **7 found**
      Select objects: *right-click*

Modifying drawings

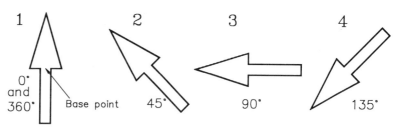

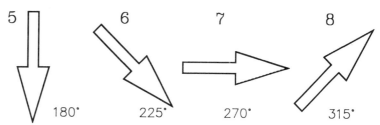

Fig. 3.15 Examples of using **ROTATE**

Base point: *pick*
<Rotation angle>/Reference: key angle in degrees   *right-click*
Command:

and the windowed objects rotate by the given angle.

If r (reference) is the response to the **<Rotation angle>/Reference:** option, two angle figures must be given to two prompts:

<Rotation angle>:/Reference: r (reference)   *right-click*
Reference <0>: key the figures of an angle   *right-click*
New angle: key the figures of the second angle   *right-click*
Command:

The first angle rotates the selected object(s) to the given angle, the second rotates the object(s) to the second angle in relation to the first.

Instead of keying an angle in response to the Rotation angle option, the rotation can be effected by movement of a rubber-band line, centered at the selected base point which appears on screen. The line can be moved around the base point under the control of the selection device. When the desired angular position of the objects being rotated appears, a *right-click* confirms that position.

The command SCALE

The whole of a drawing, or any part of it, can be altered as to its scale by using this command. Three examples of scaling a drawing

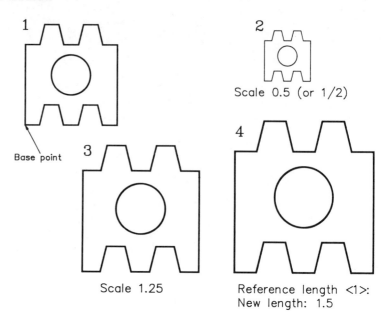

Fig. 3.16 Examples of using **SCALE**

are given in Fig. 3.16. When scaling in a CAD system, care must be taken in selecting an appropriate base point. If this selection is not made with some care, the results may not be those which were intended. When a drawing is scaled, either made smaller or larger, a badly chosen base point may have the effect of placing the scaled part in an inappropriate position in relation to other features of a drawing. Two methods of scaling are offered with this command.

1. The required scale is entered as either a fraction, such as 1/4, or as a decimal such as 0.25;
2. A reference length and a new length can be entered in response to the two prompts **Reference length:** and **New length:** which appear if r is selected in answer to the prompt <**Scale factor**>/**Reference:**.

In three of the examples of Fig. 3.16, drawings 2 and 3 are scaled to a fractional response and drawing 4 to the Reference prompts.

The **SCALE** sequences at the command line follow the pattern:

    Command: scale   *right-click*
    Select objects: w (window)   *right-click*
    First corner: *pick* or *key coordinates*
    Other corner: *pick* or *key coordinates* **21 found**
    Select objects: *right-click*
    Base point: *pick*

Modifying drawings 65

<Scale factor>/Reference: 0.5 (or 1/2)   *right-click*
Command:

or:

<Scale factor>/Reference: r (reference)   *right-click*
Reference length (1):   *right-click* to accept 1
New length: 1.5   *right-click*
Command:

The command STRETCH

Figure 3.17 shows some examples of the results of using the command. The first vertical column of drawings in Fig. 3.17 (drawings 1, 4 and 7) are of the original drawings before the command is called. The central vertical column (drawings 2, 5 and 8) demonstrate the results when the first line of drawings is stretched horizontally. The right-hand vertical column of Fig. 3.17 illustrates a vertical stretching action. The following details must be observed when the command is called into action:

1. A window MUST be used and it must be either a crossing window or a CPolygon one. Failure to observe this rule brings a warning on the command line;
2. As can be seen in drawings 5, 6 and 9, circles cannot be stretched. An arc can only be stretched if the crossing window includes an end of the arc;

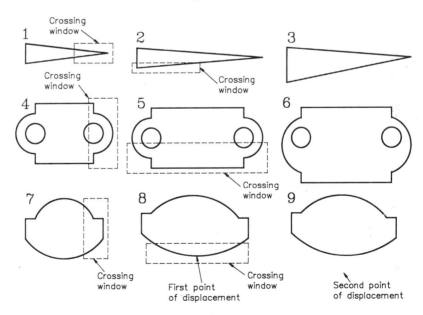

Fig. 3.17 Examples of the use of **STRETCH**

3. The command is of particular value for moving parts of a drawing within a drawing. This is illustrated in Fig. 3.18, in which several of the components in the electrical circuit have been moved with the aid of the **STRETCH** command. Drawing 4 of this illustration shows what may happen if the second point of the stretch is not in line with the connectors in the circuit. Making sure that **Ortho** is on will avoid this difficulty when moving parts such as those in the given circuit (see page 16).

The command line sequence shows the following when this command is operating:

Command: stretch   *right-click*
Select objects to stretch by window or polygon ...
Select objects: c (crossing window)   *right-click*
First Corner: *pick*   **Other corner:** *pick* **8 found**
Base point of displacement: *pick*
Second point of displacement: *pick*
Command:

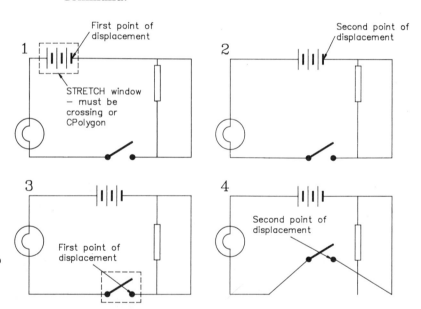

Fig. 3.18 Using **STRETCH** to move components in an electrical circuit drawing

The command TRIM

TRIM is for completing drawings constructed from a number of complete entities by "trimming" or removing those parts of the entities which are not required in the completed drawing. The command is effective on all drawing entities such as lines, arcs, circles, plines. The method of working starts by selection of the

Modifying drawings 67

entities to which entities (objects) are to be trimmed. When this selection is completed, the entities to be trimmed are then selected one after the other. As each is selected the trimming takes place. Figure 3.19 demonstrates how the command is used. The command line sequence follows the pattern:

Command: trim   *right-click*
Select cutting edge(s) ...
Select objects: *pick* **1 found**
Select objects: *pick* **1 found**
Select objects: *right-click* (to confirm that selection of cutting edges is complete)
<Select objects to trim>/Undo: *pick*
<Select objects to trim>/Undo: *pick*
<Select objects to trim>/Undo: *pick*
<Select objects to trim>/Undo: *right-click* (to confirm that trimming is ended)
Command:

Drawings 4 and 5 of Fig. 3.19 show a typical example where trimming is an effective and speedy method of constructing a required outline. Drawings 1 to 7 of Fig. 3.20 illustrate the sequence of trimming objects to obtain the outline of a simple tool handle from a set of tangential circles.

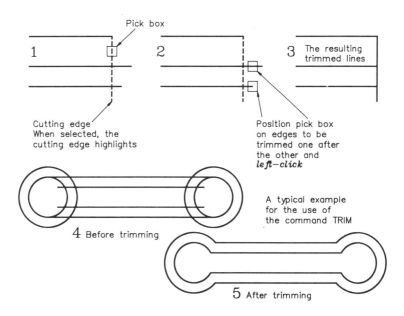

Fig. 3.19 Examples of the use of **TRIM**

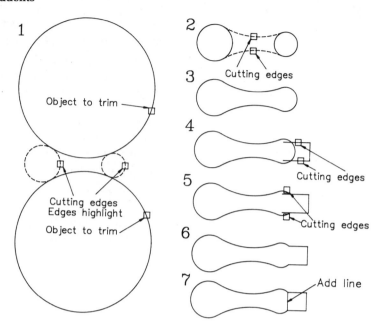

Fig. 3.20 Further examples of the use of **TRIM**

The command BREAK

With **BREAK**, gaps can be made in entities or they can be broken in such a manner as to cause gaps to appear in their length or outline. If the command is selected from the **Modify** pull-down menu, three options are offered:

>Select Object, 2nd Point;
>Select Object, Two Points;
>At Selected Point.

When keying the command at the command line, these options become prompts showing as:

>**Command:** break   *right-click*
>**BREAK Select object:**   *pick*
>**Enter second point (or F for first point):**   *pick*
>**Command:**

and the entity breaks as in drawing 2 of Fig. 3.21.

If a two-point option is required the command line shows:

>**Command:** break   *right-click*
>**BREAK Select object:**   *pick*
>**Enter second point (or F for first point):** f (first)   *right-click*
>**Enter first point:**   *pick*

Enter second point: *pick*
Command:

and the line breaks as in drawing 4.

If, in response to the **Enter second point (or F for first point):** prompt, an @ is entered, or in response to the **Enter second point:** prompt, the same point is picked as for the first, the entity is broken at that point. Drawings 5 and 6 of Fig. 3.21 show the result of these two responses.

Figure 3.22 shows the results of using the **BREAK** command on arcs, circles and ellipses.

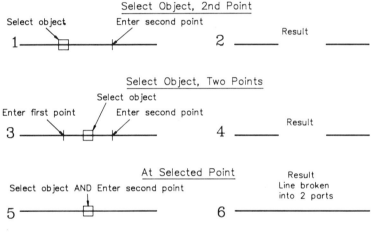

Fig. 3.21 The use of the command **BREAK**

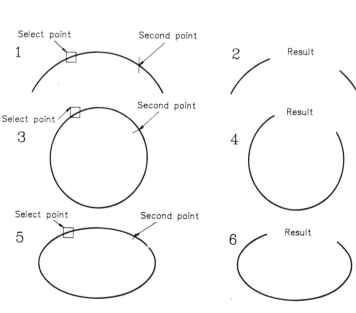

Fig. 3.22 **BREAK** used with arcs, circles and ellipses

The command EXTEND

Figure 3.23 demonstrates the action of the command **EXTEND**. The command can be used to extend an entity onto another entity. After selecting the entity onto which the extension is to take place, the entity to be extended is selected and the extension takes place. The command line shows the following when the command is in action:

Command: extend *right-click*
Select boundary edges ...
Select objects: *pick* **1 found**
Select boundary edges ...
Select objects: *pick* **1 found**
Select boundary edges ...
Select objects: *right-click*
<Select object to extend>/UNDO: *pick*
<Select object to extend>/UNDO: *pick*
<Select object to extend>/UNDO: *right-click*
Command:

In the same way as lines or plines can be extended as shown in Fig. 3.24, so can arcs, or lines, plines or arcs be extended to lines, plines, arcs or circles.

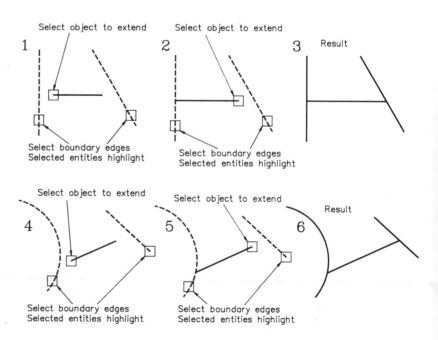

Fig. 3.23 Examples of the use of **EXTEND**

Modifying drawings

The command PEDIT

Only a simplified explanation of this command is given here because it is doubtful if the reader learning the use of AutoCAD or the student reader will use this command to any extent. The command is for editing polyline entities. When called, the command line shows:

Command: pedit   *right-click*
Select polyline: *pick*
**Close/Join/Width/Edit vertex/Fit/Spline/Decurve/Ltype gen/
Undo/eXit (X):**

The results of selecting three of these prompts are indicated in Fig. 3.24. In drawing 1, a pline width is changed by the **Width** option; in drawing 2, a pline arc is straightened by the **Decurve** option; in drawing 3 one of the vertices of a pline is moved with the **Edit vertex** option.

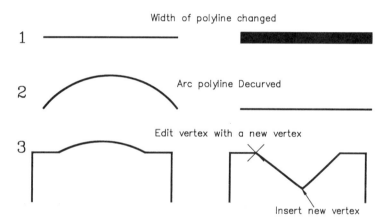

Fig. 3.24 Examples of the use of **PEDIT**

Revision notes

1. As with all other commands in AutoCAD, when the **Modify** commands are called, no matter whether keyed or picked from a menu, the command name and the options available with the command appear in the on-screen menu area;
2. The prompts connected with many commands in AutoCAD, particularly those from the **Modify** menu, show the term **Object** – this has the same meaning as the term **Entity**;
3. When commands from the **Modify** menu are called, any group of objects can be placed in a window without responding with the w (window) option. Thus, without keying w, *pick* a point

on screen. This is treated as a **First corner** of a window and the **Other corner:** prompt appears at the command line. In the examples given above it has been assumed that a w must be called;

4. It is the commands in the **Modify** menu, together with commands such as **COPY**, that give rise to the CAD saying:

   Never draw the same thing twice;

5. Each of the command systems in the **Modify** menu has a similar set of prompts and options;
6. As with other angle actions in AutoCAD angular rotation is anti-clockwise, although a clockwise direction can be configured if desired;
7. A rule for the selection of commands which many operators follow is to use keyboard abbreviations for commands where they are available, but to select commands from pull-down or on-screen menus when an abbreviation is not available. The reason for this is that it may be quicker to select than to key a full command name, whereas it is probably quicker to key an abbreviation than to select;
8. The **Change** and **Explode** commands from the menu will be dealt with in a later chapter. Some details of drawings which can be changed have not so far been mentioned.

Exercises

Exercise 1

Figure 3.25 (top part). Construct the dimensioned outline. Then:

1. Multiple COPY the outline six times and then:
2. ROTATE each of the copies as in Fig. 3.25.

Do not include any dimensions in your answer.

Exercise 2

Figure 3.26. This drawing can be constructed either by drawing a series of short lines or by drawing lines from side to side and from top to bottom to the sizes indicated and then to TRIM unwanted parts of lines. The second method using TRIM is quicker.

Do not dimension your answer.

Modifying drawings

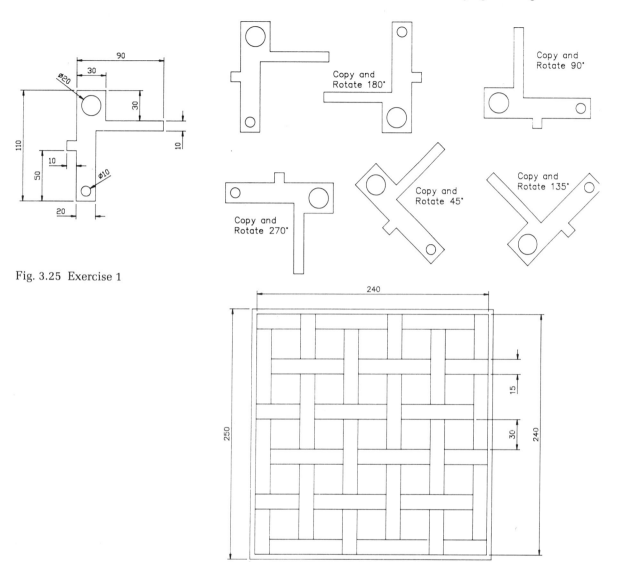

Fig. 3.25 Exercise 1

Fig. 3.26 Exercise 2

Exercise 3

Figure 3.27. Construct only the outline of the house to the dimensions given. Then:

1. Add the door;
2. Construct one window;
3. Multiple COPY your window to the other required positions.

Do not include dimensions.

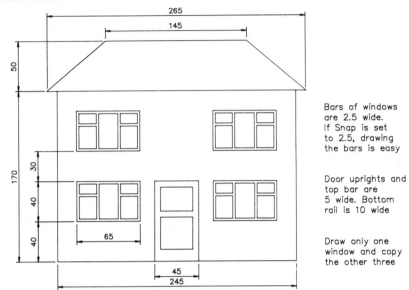

Fig. 3.27 Exercise 3

Bars of windows are 2.5 wide. If Snap is set to 2.5, drawing the bars is easy

Door uprights and top bar are 5 wide. Bottom rail is 10 wide

Draw only one window and copy the other three

Exercise 4

Figure 3.28. Construct the wheel elevation with CIRCLE, PLINE and ARRAY.

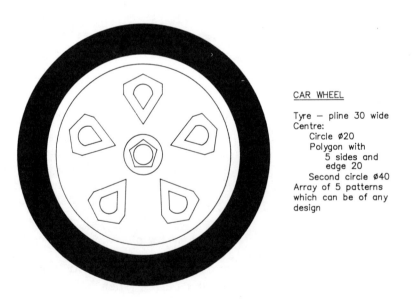

CAR WHEEL

Tyre — pline 30 wide
Centre:
    Circle ⌀20
    Polygon with
        5 sides and
        edge 20
    Second circle ⌀40
Array of 5 patterns
which can be of any
design

Fig. 3.28 Exercise 4

Exercise 5

Figure 3.29. With LINE, CIRCLE and TRIM construct the "maze" to the given dimensions. Do not include the dimensions.

Modifying drawings 75

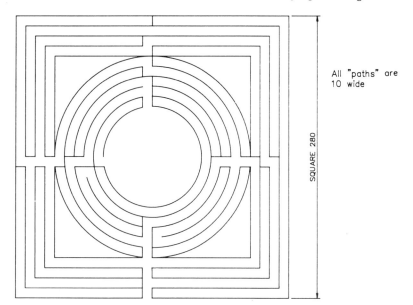

Fig. 3.29 Exercise 5

CHAPTER 4

Layers

Introduction

The drawing of different forms of graphic features such as outlines, centre lines, hidden detail lines, text, etc., each on its own layer is common to practically all CAD software packages. Other features such as the drawing of each floor in a multi-storey building, or each view of a multi-view projection, can be constructed on its own layer. Each layer can be assigned to its own colour and its own linetype. The advantage of working on layers becomes most apparent when constructing complicated drawings. When, for example, constructing the outlines of a front view of an intricate machined component, if all other views are constructed, each on its own layer, only that layer on which the front view has been drawn need be ON, all others can be turned OFF.

The options in the LAYER command

The command can be brought into operation by any one of the following:

1. Entering la (layer) at the command line from the keyboard;
2. Selecting **Layer Control ...** from the **Settings** pull-down menu;
3. Selecting **Layer ...** from the on-screen menu.

Of these three methods, choosing either of those which bring the **Layer Control** dialogue box on screen is possibly the easiest to use. When selecting from either the **Settings** pull-down menu or from **LAYER ...** in the on-screen menu, the **Layer Control** dialogue box (Fig. 4.1) appears on screen and the settings for any layer already in the drawing can be chosen from this dialogue box, or new layers can be made and their settings determined.

When la (layer) is entered at the keyboard, the layer settings options appear at the command line:

Command: la (layer) *right-click*
**?/Make/Set/New/ON/OFF/Color/Ltype/Freeze/Thaw/Lock/
Unlock:**

The options provide the following:

? brings a list of layers in the current drawing on screen, together with their linetypes and colours;
Make: enter m (make) at the keyboard in order to make a new layer;
Set: enter s (set) to set an existing layer as the current layer;
New: enter n (new) to add a new layer to those already in the drawing under construction;
ON: enter on to turn on a layer which has been off;
OFF: enter off to turn off a layer;
Color: enter c (color) to change the colour of the entities on a layer;
Ltype: enter l (ltype) to change the type of line in use on a layer;
Freeze: enter f (freeze) to freeze a layer. The frozen layer is not only turned OFF but, when a drawing is regenerated as in a ZOOM, the layer is not regenerated, with the result that the regeneration is quicker. Can save considerable time when constructing a complicated drawing;
Thaw: enter t (thaw) to unfreeze a frozen layer;
Lock: enter l (lock) to lock a layer. Entities on a locked layer can be seen, but cannot be edited (e.g. erased, moved, copied, etc.), but new entities can be added to the layer. Of greatest value when constructing a complicated drawing because it prevents editing by mistake of the entities on a locked layer;
Unlock: enter u (unlock) to unlock a locked layer.

The Layer Control dialogue box

These options are all available and can be set from the **Layer Control** dialogue box (Fig. 4.1). *Left-click* after positioning the arrow cursor onto a layer name and the name highlights. The highlighted layer can then be turned On or Off, Frozen, Thawed, Locked or Unlocked by a *left-click* on the appropriate button. In the same manner the highlighted layer can be made as the Current layer, as a New layer or Renamed. Two other dialogue boxes associated with the settings are the **Select Linetype** (Fig. 4.2) and the **Select Color** (Fig. 4.3). Settings are made from these dialogue boxes by a *left-click* on the appropriate linetype or colour in the boxes, followed by a *left-click* on the **OK**.

78   AutoCAD Release 12 for students

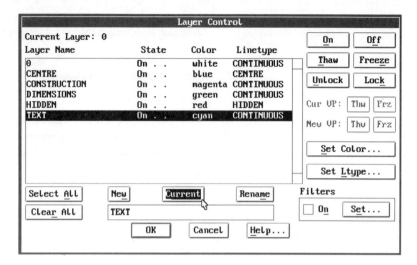

Fig. 4.1 The **Layer Control** dialogue box

Linetypes

A large number of different types of line are available in AutoCAD 12. Any one of these can be loaded for use on a layer, set as the current linetype or new linetypes can be created by the operator. Creation of a new linetype will not be dealt with here. For example, to load the linetype centre:

      Command: linetype   *right-click*
      ? Set/Load/Create: l (load)   *right-click*
      Linetype(s) to load: centre   *right-click* and the Linetype
              File dialogue box appears   *Return* (not *right-click*)
      Linetype CENTRE loaded.
      ? Set/Load/Create:
      Command:

Notes

1. If a ? is entered as a response, a screen appears showing the linetypes available in the acad/support/acad.lin file.
2. If the **Select Linetype** dialogue box is called by a *left-click* on the **Set Ltype** box in the **Layer Control** dialogue box, only the linetypes already loaded will show in the dialogue box (Fig. 4.2).

Layers 79

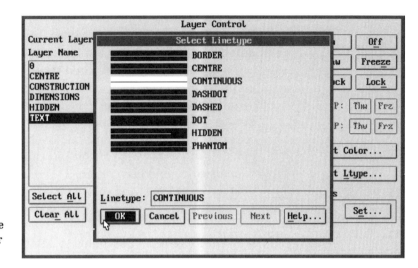

Fig. 4.2 The **Select Linetype** dialogue box from the **Layer Control** dialogue box

Setting colours

The layer colour can be set from the **Layer Control** dialogue box by a *left-click* on **Set Color ....** This brings up the **Select Color** dialogue box (Fig. 4.3). A *left-click* on the appropriate colour in this box, followed by a *left-click* on **OK** sets the colour for the layer. The colour can also be set after entering c (color) as a response to the prompts:

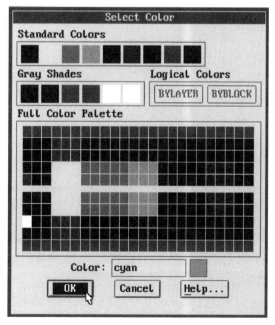

Fig. 4.3 The **Select Color** dialogue box from the **Layer Control** dialogue box

?/Make/Set/New/ON/OFF/Color/Ltype/Freeze/Thaw/Lock/
Unlock:

followed by entering the colour name (e.g. red) or the colour number (e.g. for red use the figure 1).

If a colour change is required when adding an entity or group of entities on a layer in a colour different to the layer colour, the command **COLOR** can be called and either the colour name or its number entered as a response to the prompts appearing with the command. However, it must be noted that the colour so called will remain as that in which all further construction will appear until it is changed to another colour.

The command CHANGE

It is appropriate at this point to include details of this command because it may involve a layer. When called the command line shows:

>**Command:** change   *right-click*
>**Select objects:**   *pick* an entity, e.g. a continuous line
>**Select objects:**   *right-click*
>**Properties/<Change point>:** p (properties)   *right-click*
>**Change what property (Color/Elev/LAyer/LType/Thickness)
>? lt (ltype)   *right-click*
>**New linetype <BY LAYER>:** centre   *right-click*
>**Change what property (Color/Elev/LAyer/LType/Thickness)
>?   *right-click*
>**Command:**

and the entity changes from a continuous line to a centre line.

Other drawing details can be changed as will be seen in the above prompts – Color, Elevation (see page 170), Layer and Thickness (associated with Elevation).

If changing the style of text already on screen, when the command line of the change sequence shows:

>**Properties/<Change point>:**   *pick* the text to be changed
>**New Style or RETURN for no change:** romanc   *right-click*
>(see Fig. 4.4)
>**New rotation angle:**   *right-click*
>**New text<SIMPLEX<12 high>:**   *right-click*
>**Command:**

and the text style is changed from Simplex to Romanc.

Some changes effected by the command are shown in Fig. 4.4.

Layers

A CONTINUOUS line changed to a CENTRE line

SIMPLEX text changed to ROMANC text

Fig. 4.4 Examples of the use of **CHANGE**

A CONTINUOUS Pline changed to a HIDDEN detail Pline

Notes on layers

Figure 4.5 is a simple engineering drawing such as might have been drawn by a student learning how to use AutoCAD 12. Figure 4.6 shows, in diagrammatic form, the six layers on which the drawing was constructed.

Notes

1. The operator can construct his/her drawing on as many layers as thought fit, there being no upper limit.
2. Layers could be regarded as being similar to a series of tracings one on top of the other. As with tracings, one or more can be removed from the series – compare with the turning off layers by **OFF**. A tracing which has been removed can be replaced – compare with turning a layer **ON**.
3. When a new drawing is called, the current layer will always be Layer 0;
4. Layer 0 cannot be removed from the drawing, although it can be **OFF**, Locked or Frozen.

A work disk

Reference was made earlier (page 18) to a work disk (*a:\work.dwg*) to which examples and exercises from this book could be saved as AutoCAD drawing (*.dwg*) files. The limits of the drawing area for that file were set to 420,297 – A3 sheet size (in millimetres). The work file could now be updated to include a set of layers such as those shown in Fig. 4.6:

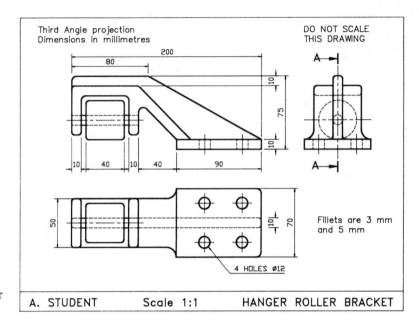

Fig. 4.5 An orthographic projection of a Hanger Roller Bracket

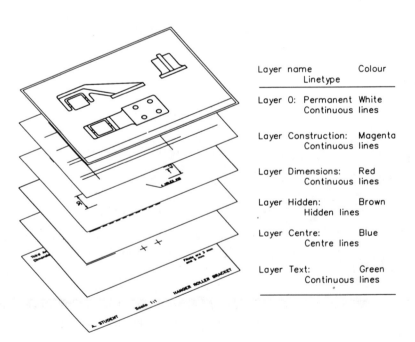

Fig. 4.6 The layers involved in the drawing Fig. 4.5

Layer name	Colour	Linetype
0	White	Continuous
Construction	Magenta	Continuous
Dimensions	Red	Continuous
Hidden	Brown	Hidden
Centre	Blue	Centre
Text	Green	Continuous

An AutoCAD 12 example drawing

Figures 4.7 and 4.8 are examples of AutoCAD 12 constructed on layers. They show two sheets of drawings from a number of sheets for the design of a small table lectern. These two illustrations are included here to show the reader that AutoCAD can be used for any form of technical drawing.

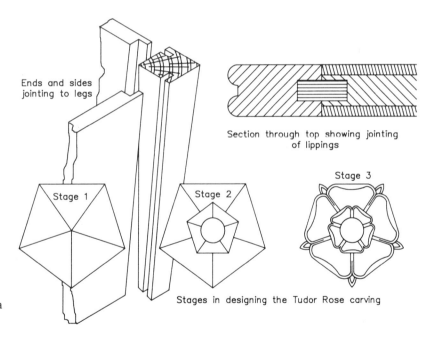

Fig. 4.7 One of several design sheets for designing a table lectern

84 AutoCAD Release 12 for students

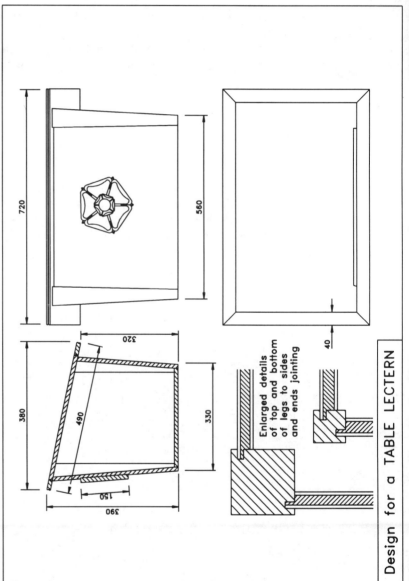

Fig. 4.8 An orthographic projection of the table lectern with some details of its construction

CHAPTER 5

The Construct pull-down menu

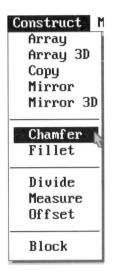

Fig. 5.1 The **Construct** pull-down menu

Introduction

The commands in the **Construct** pull-down menu (Fig. 5.1) are scattered through the various on-screen menus. One of these commands – **Copy** – has already been described in Chapter 3 because its command sequences are so similar to those of the command **Move** – found in the **Modify** pull-down menu.

The command MIRROR

The command **Mirror** is of good value when constructing drawings which are symmetrical about one or more axes. With such drawings, the command saves the effort in repeating the construction of those parts which are identical, but which are a mirror-image of parts already constructed. In technical drawings of all types, such drawings are quite common. For this reason it is advisable for the reader to practise using the command.

Figure 5.2 is an example of an outline, which being symmetrical

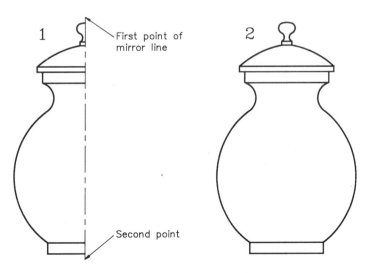

Fig. 5.2 An example of using **MIRROR**

around a vertical axis is suitable for mirroring. Drawing 1 is the half-outline before the command is in operation; drawing 2 shows the result of the action of the **MIRROR** command. The command line sequences for this example follow the pattern:

Command: mirror   right-click
Select objects: w (window)   right-click
First corner: pick   **Other corner:**   pick **8 found**
Select objects: right-click
First point on mirror line: pick   **Second point:**   pick
Delete old objects <N>:   right-click
Command:

and the mirror image appears on screen. If yes is keyed to **Delete old objects <N>**, only the right-hand half of drawing 1 would have appeared on screen.

A second example is given in Fig. 5.3. This example is symmetrical around two axes. Thus only a quarter of the drawing needed to be constructed before mirroring twice to obtain the whole drawing. The quarter part was constructed (drawing 1); that part was then mirrored below the horizontal axis; the half drawing was then mirrored around its vertical axis (drawing 2) to produce the completed full drawing (drawing 3). Finally the central hole was added, together with its centre lines.

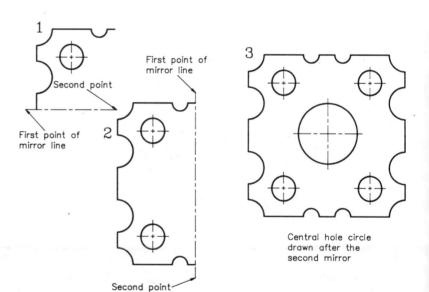

Fig. 5.3 A second example of using **MIRROR**

A special case arises when text of any kind is mirrored. The mirror image of text will obviously be with all letters back-to-front or upside down. AutoCAD 12 deals with this problem with the aid of the set variable **MIRRTEXT**. This variable can be set to either 0 or 1 by:

Command: mirrtext *right-click*
New value for MIRRTEXT <1>: 0
Command:

and any mirrored text is automatically placed on screen the right way round – i.e. in a properly readable form.

Examples of text mirrored with the variable set at both 1 and 0 are given in Fig. 5.4.

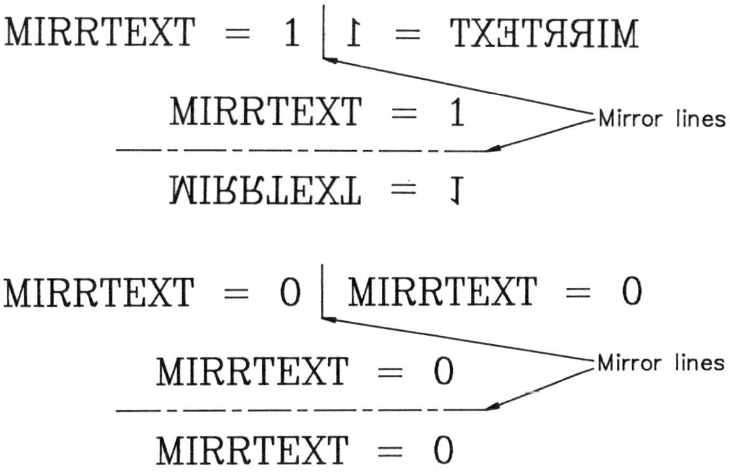

MIRRTEXT = 1 | 1 = TXETRRIM

MIRRTEXT = 1 ⟩Mirror lines
─ ─ ─ ─ ─ ─ ─ ─
WIKKLEXL = I

MIRRTEXT = 0 | MIRRTEXT = 0

MIRRTEXT = 0 ⟩Mirror lines
─ ─ ─ ─ ─ ─ ─ ─
MIRRTEXT = 0

Fig. 5.4 The value of using **MIRRTEXT**

The command ARRAY

ARRAY is used to produce multiple copies of a drawing, which are either in rows and columns or rotated around a selected central point. Several examples of drawings produced with the aid of this command are given in Figs 5.5 and 5.6. The two types of arrays are Rectangular (rows and columns of copies) and Polar (copies around a central point). Note the following rules governing the answers to prompts in the command structure.

Rectangular arrays

1. (a) Spacing between rows are stated in coordinate units vertically:
      distances vertically upwards are +ve;

distances vertically downwards are −ve;
These two rules correspond to +ve and −ve values of the y coordinate;
(b) Spacing between columns are stated in coordinate units horizontally;
distances to the right are +ve;
distances to the left are −ve;
These two rules correspond to +ve and −ve values of the x coordinate;
2. Spacings between rows and columns can be set by picking points on the screen which are the required distance apart, instead of keying in figures;
3. Assume that the spacing figures are taken from the bottom left-hand corner of the part being arrayed. This may not necessarily be fully correct, but it is a good rule to follow if one is to avoid the possible problem of the arrayed copies overlapping each other.

Taking drawings 1 and 2 of Fig. 5.5 as an example, the sequence of prompts for a Rectangular array are:

Command: array   right-click
Select objects: w (window)   right-click
First corner: pick   **Other corner:** pick **4 found**
Rectangular or Polar array (R/P) <P>: r (rectangular)
right-click

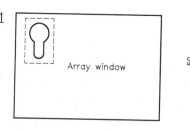
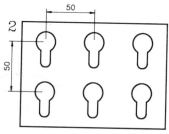
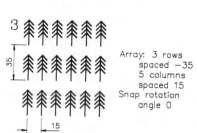
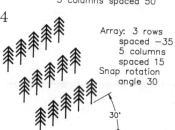

Fig. 5.5 Examples of rectangular **ARRAY**s

> **Number of Rows (—)<1>:** 3   *right-click*
> **Number of Columns (|||)<1>:** 2   *right-click*
> **Unit cell or distance between rows (—):** −50   *right-click*
> **Distance between columns (|||):** 50   *right-click*
> **Command:**

Rectangular arrays can be produced at an angle to the x,y coordinate axes by setting **Snap** to a rotation angle. To do this call **Snap** and follow a sequence such as:

> **Command:** snap   *right-click*
> **Snap spacing or ON/OFF/Aspect/Rotate/Style<5>:** r
>   (rotate)   *right-click*
> **Base point <0,0>:**   *left-click* to accept (0,0)
> **Rotation angle <0>:** 30   *right-click*
> **Command:**

If **Grid** is on, the screen grid dots will now be set at the snap angle. Rectangular arrays will also be the **Snap** angle. Drawing 4 of Fig. 5.5 is an example of a rectangular array with **Snap** set at 30 degrees.

Polar arrays

1. The rotation of the array may be either clockwise (cw) or anticlockwise (ccw). When completing an array within a whole circle (360 degrees), it usually does not matter whether the array is cw or ccw, but occasions may arise when the direction of rotation of the array as the part in copies may be important;
2. Copies can either be rotated as they are copied or not rotated. If rotated the rotation will be to an angle with respect to the centre point of the array. If not rotated, remember the copies are with respect to the selected centre point.

The sequence of prompts at the command line for a Polar array is:

> **Command:** array   *right-click*
> **Select objects:** w (window)   *right-click*
> **First corner:** *pick*   **Other corner:** *pick* **4 found**
> **Rectangular or Polar array (R/P) <R>:** p (polar)   *right-click*
>
> **Center point of array:** *pick*
> **Number of items:** 7   *right-click*
> **Angle to fill (+=ccw,−=cw)<360>:**   *right-click* to accept 360

Rotate objects as they are copied? <Y>: *right-click to accept YES*
Command:

and the object(s) are copied around the centre point. Figure 5.6 gives two examples of Polar arrays formed in this manner.

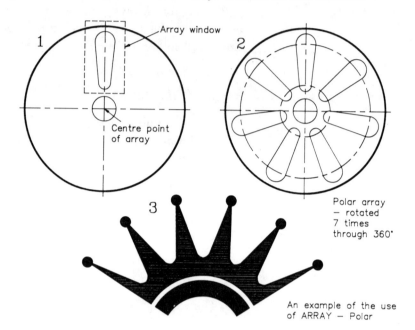

Fig. 5.6 Examples of using Polar **ARRAY**s

The command CHAMFER

When the command chamfer is entered at the keyboard, the command line shows:

Command: chamfer   *right-click*
Polyline/Distance/<Select first line>: d (distance)   *right-click*

Enter first chamfer distance <0>: 20   *right-click*
Enter second chamfer distance <20>:   *right-click* (to accept 20)
Command: *right-click*
CHAMFER Polyline/Distance/<Select first line>:   *pick*
Select second line: *pick*
Command:

and the chamfer automatically forms – Fig. 5.7, drawings 1 and 2.
The usual first response to the **Polyline/Distance/<Select first line>:** options is d (for Distance) – in order to set the sizes for the

The Construct pull-down menu

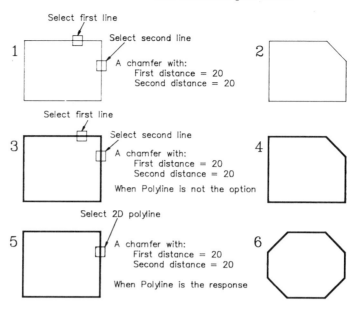

Fig. 5.7 The command
CHAMFER

chamfer. With polylines there are two methods for constructing a chamfer, depending upon whether each corner of the pline is to be chamfered or only one corner. Drawings 3 and 4 of Fig. 5.7 show a chamfer formed when the pline is treated as if it were a line – i.e. the **Polyline** option in the set of prompts associated with the **Chamfer** command is not called. Drawings 5 and 6 of Fig. 5.7 show the result when a p (for Polyline) is the response to the first prompt at the command line.

If the response to the set of prompts is a *right-click*, the distances set for the last chamfer to be drawn will be the current distances. Not until a new pair of distances are set will the current distances be changed.

The command FILLET

The actions associated with this command are very similar to those for the command **Chamfer**, except that corners are "filleted" – i.e. radiused – and only one distance (a radius) is required to set the current size of the fillet. The prompts and responses at the command line are:

Command: fillet   *right-click*
Polyline/Radius/<Select first object>: r (radius)   *right-click*

Enter fillet radius<0>: 20   right-click
Command: right-click
FILLET Polyline/Radius/<Select first object>:   pick
Select second object: pick
Command:

and the fillet forms, to the given radius at the junction of the two picked objects – usually lines.

Note the similarity of action when polylines are being filleted, as indicated in Fig. 5.8. If the response to the first set of options is p (for Polyline), then all corners of the pline are filleted after only one *pick* point has been chosen anywhere on the plines. Drawings 3 to 6 show the effects of either not responding with a p (Polyline) or responding with a p.

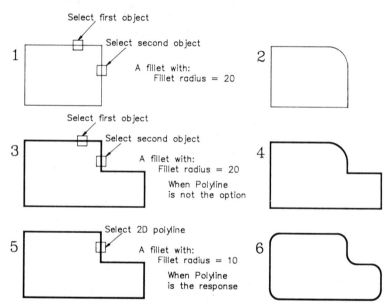

Fig. 5.8 The command **FILLET**

The command OFFSET

With the aid of this command entities can be drawn which are parallel along their length to a selected object. Figure 5.9 shows the effect of using the command on the most commonly used construction entities in an AutoCAD drawing. Enter the command and the command line shows:

Command: offset   right-click
Offset distance or Through <Through>: 10   right-click
Select object to offset: pick

Side to offset? *pick*
Select object to offset: *pick*
Side to offset? *pick*
Select object to offset: *right-click*
Command:

and the offset entities appear. If the first offset object is picked a second offset appears. If this offset is then picked a third offset appears and so on until a *right-click* is given, when the command line will revert to **Command:** ready for the next keyboard entry.

Figure 5.9 shows the effect of using Fillet and Offset both on the same outlines.

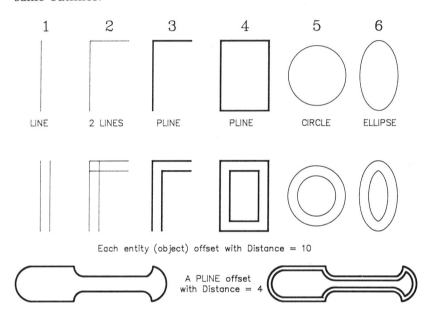

Fig. 5.9 The command **OFFSET**

Hatching

Two methods of hatching are available with AutoCAD 12:

1. The method used in earlier releases in which each entity of the boundary of a hatched area must be selected by *picking* before the hatching can be applied. This method is operated by entering commands and responses from the keyboard;
2. A method introduced with Release 12 called **BHATCH**, which, together with several dialogue boxes, provides a quicker method and also one which allows a preview of the hatch before it is applied in order for the operator to check whether the hatching is as wished.

The command BHATCH

Taking as an example the hatching for the TOOL POST in Fig. 5.10. When bhatch is entered at the keyboard the following sequence needs to be followed in order to produce a successful hatched area:

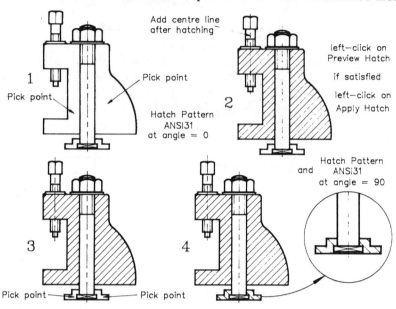

Fig. 5.10 An example of using **BHATCH**

    Command: bhatch   right-click
      The *Boundary Hatch* dialogue box comes on screen
                                                   (Fig. 5.13)
    left-click on **Hatch Options ...**
    The **Hatch Options** dialogue box comes on screen
                                                     (Fig. 5.14)
    left-click on Select **Pattern ...**
    The **Hatch Pattern** dialogue box comes on screen
                                                     (Fig. 5.15)
    left-click on the box with the **ansi31** pattern
    The **Hatch Options** dialogue box re-appears
    Enter e.g. 1.5 in the **Scale:** box
    left-click on **OK**
    The **Boundary Hatch** dialogue box re-appears
    left-click on the **Pick Points** box
Select internal point   *pick*
Selecting everything visible ...
Analysing the selected data ...
Select internal point   *pick*
Select internal point   *right-click*

The **Boundary Hatch** dialogue box re-appears
left-click on the **Preview Hatch** <box

the selected picked area hatches. If satisfied that the hatch is as desired:

right-click and the **Boundary Hatch** dialogue box re-appears

left-click on the **Apply Hatch** box

the hatching is applied in the area in the drawing.

This all seems to take a long time, but it is, in fact, a speedy method for hatching areas. The reading of the example above probably takes much longer than the hatching once the method is understood and followed.

Figure 5.11 is another example of hatching applied with the aid of the command **BHATCH**. Figure 5.12 shows six examples of different hatch patterns obtained from the numerous different hatch patterns in the **Hatch Pattern** dialogue.

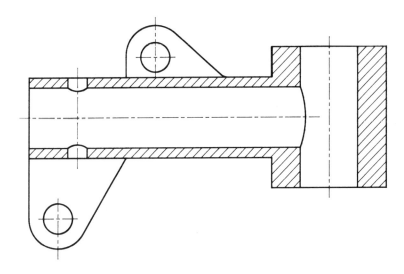

Fig. 5.11 Another example of **BHATCH**

Notes on the command BHATCH

1. **BHATCH** provides an easy-to-use hatching method under the control of selections from dialogue boxes;
2. **BHATCH** provides the opportunity of pre-viewing hatched areas and to amend the hatching pattern, size and angle before the hatching is finally added to a drawing;
3. The angle of hatching is relevant to the angle at which the hatch pattern appears as an icon in the **Hatch Pattern** dialogue

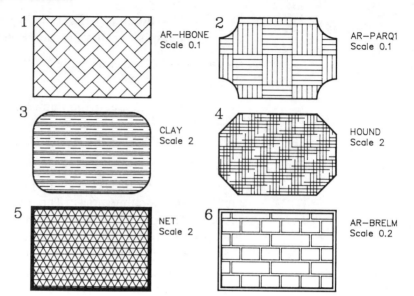

Fig. 5.12 Six hatch patterns from those in the **Hatch Patterns** dialogue box

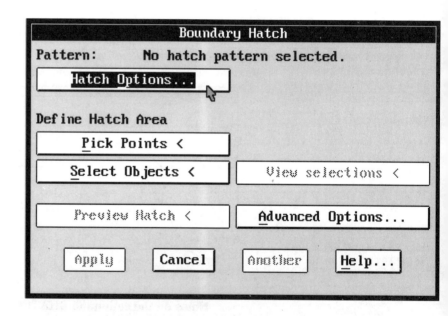

Fig. 5.13 The **Boundary Hatch** dialogue box

   box. The icon is assumed as being displayed at the angle 0;
4. The **Normal** style of hatching is that most frequently used when deciding upon selecting the style from the **Hatching Style** box in the **Hatch Options** dialogue box (Fig. 5.15);
5. The hatch patterns showing in the **Hatch Pattern** dialogue box are contained in the AutoCAD 12 file *acad.pat*;
6. Hatching can only be correctly applied within a closed

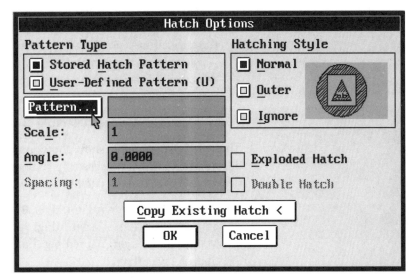

Fig. 5.14 The **Hatch Options** dialogue box

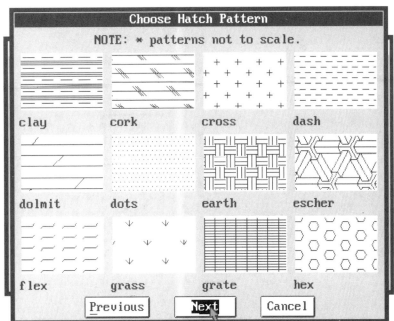

Fig. 5.15 One page of the **Choose Hatch Pattern** dialogue box

boundary. If the slightest gap exists in the boundary the hatching will *leak* over the drawing;
7. Hatching can be exploded if necessary by *picking* any point within a hatched area after calling the command **EXPLODE**;
8. A hatched area can be erased by calling **ERASE**, *picking* any point inside the area followed by a *right-click*. This is because

hatched areas are added to a drawing as blocks — see page 135;
9. If the command **HATCH** is called, hatching is applied following the procedures previously used with earlier AutoCAD releases — each boundary entity of an area to be hatched must be selected before hatching can be applied.

A more advanced example of hatching

Figures 5.16 to 5.19 show how areas within parts of drawing boundaries can be successfully hatched. The hatching of the final drawing Fig. 5.19 was carried out by following the sequence:

1. Construct the drawing on layer **0** – Fig. 5.16 – in this example an end elevation of a building;
2. Make a new layer, named e.g. **hatch**, with a colour different to that of layer **0**;
3. Make layer **hatch** current;

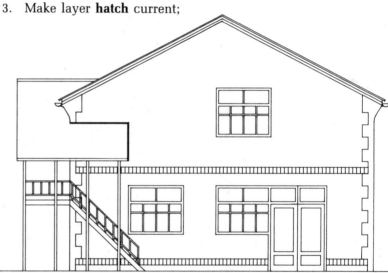

Fig. 5.16 The building elevation on layer **0** before hatching

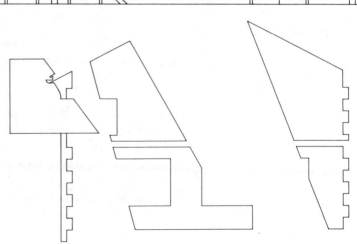

Fig. 5.17 The hatch boundaries constructed on layer hatch

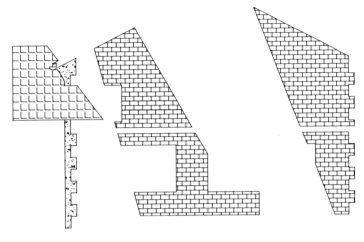

Fig. 5.18 Hatching applied within boundaries on layer **hatch01**

Fig. 5.19 The completed elevation with all hatching completed

4. With the elevation still on screen and with the aid of **Snap** and the **Osnaps** carefully draw the outlines of the areas to be hatched – Fig. 5.17;
5. Make a new layer – e.g. named **hatch01** – preferably of a different colour to both layers **0** and **hatch** and make layer **hatch01** current;
6. Turn layer **0** off;
7. With the aid of **BHATCH** and its dialogue boxes, hatch each of the areas previously outlined on layer hatch – Fig. 5.18;
8. When satisfied with the hatch areas as to pattern, scale and angle, turn layer **0** on and make it the current layer and turn layer **hatch** off;
9. The resulting completed drawing is shown in Fig. 5.19.

Exercises (some revision)

Exercise 1

There are five drawings in Fig. 5.20 for this exercise.

Drawing 1 is just a piece of fun. Construct the "face";

Drawing 2 – construct the half-drawing which is dimensioned and then MIRROR your half-drawing to produce the complete construction;

Drawing 3 – draw the six circles;

Drawing 4 – TRIM parts of the six circles;

Drawing 5 – SCALE drawing 4 to twice full size.

Do not include dimensions in your answers.

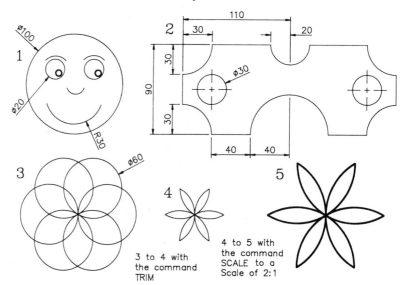

Fig. 5.20 Exercise 1

Exercise 2

Figure 5.21. Draw the circle and the upper nut plan (POLYGON and CIRCLE). Then Polar ARRAY the single nut 8 times around the full 360 degrees.

Exercise 3

Figure 5.22. Construct drawing 1 without including any dimensions. Then follow the instructions given with Fig. 5.22.

The Construct pull-down menu

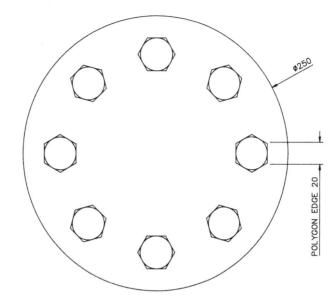

Fig. 5.21 Exercise 2

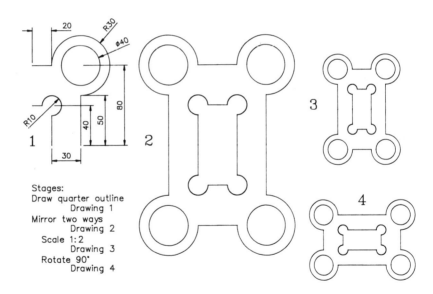

Fig. 5.22 Exercise 3

Exercise 4

Figure 5.23. This exercise involves your answer to Exercise 3. Follow the instructions given with Fig. 5.23.

Dimensions should not be included in your answer.

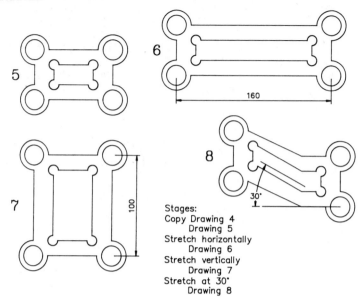

Fig. 5.23 Exercise 4

Stages:
Copy Drawing 4
　　Drawing 5
Stretch horizontally
　　Drawing 6
Stretch vertically
　　Drawing 7
Stretch at 30°
　　Drawing 8

Exercise 5

Figure 5.24 is the outline of a sectional view through a drill tray from a drilling machine. Copy the given outline and add hatching as indicated.

Do not include the dimensions.

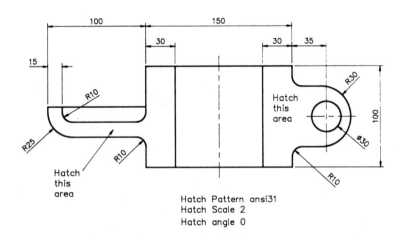

Hatch Pattern ansi31
Hatch Scale 2
Hatch angle 0

Fig. 5.24 Exercise 5

Exercise 6

Copy Fig. 5.25 without the dimensions.

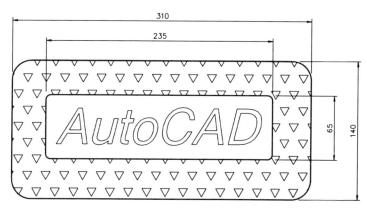

Fig. 5.25 Exercise 6

Exercise 7

The outline of a sectional view through a coupling gland is given in Fig. 5.26. Copy the given outline and add hatching as indicated. Do not dimension your drawing.

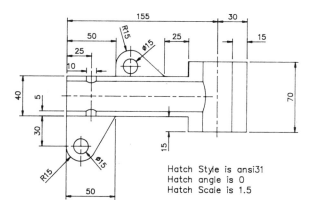

Fig. 5.26 Exercise 7

Exercise 8

Figure 5.27 is an end elevation of a house. Draw the house to the sizes and hatchings given with the drawing but do not include the dimensions.

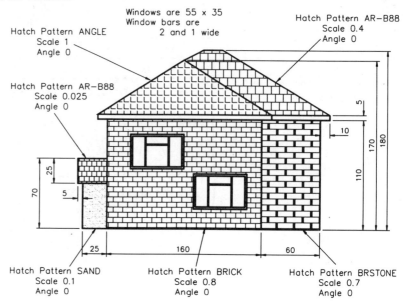

Fig. 5.27 Exercise 8

Exercise 9

Another revision exercise. Draw the outline of any saloon car such as that shown in Fig. 5.28.

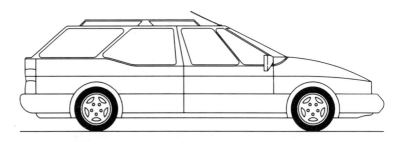

Fig. 5.28 Exercise 9

A note on text in hatched areas

Text is surrounded by an invisible boundary which hatching will not pass through providing the hatch boundaries and the text are selected as separate entities. This is achieved by a *left-click* on the **Select Objects** box in the **Boundary Hatch** dialogue box (see Fig. 5.13). This must be followed by selecting each object (entity) of the boundary to the area to be hatched and also selecting each item of text. An example is given in Fig. 5.29. In this example the two Hatch Patterns are **ansi37** and **honey**, both at a scale of 1 and an angle of 0.

Fig. 5.29 An example of hatching that includes text

CHAPTER 6

The Settings menu. Dimensioning

Introduction

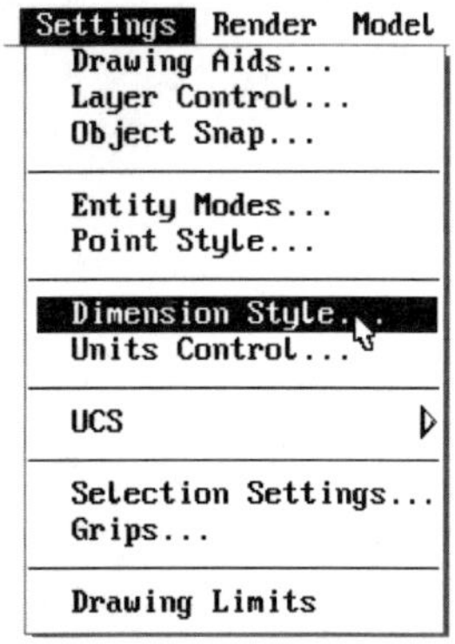

Sufficient details have already been given in earlier pages about most of the dialogue boxes which can be called by selection from the **Settings** pull-down menu. Figure 6.1 shows the contents of this menu. The pages on which references to these dialogue boxes can be found, together with the command name which can be entered from the keyboard instead of selecting the dialogue box name from the **Settings** menu are:

Drawing Aids ... page 15. Called by entering ddrmodes;
Layer Control ... page 78. Called by entering ddlmodes;
Object Snap ... page 44. Called by entering ddosnap;
Entity Modes ... page 57. Called by entering ddemodes;
Point Style ... page 57. Called by entering ddptype;
Dimension Style ... the main topic of this chapter;
Units Control ... page 15. Called by entering ddunits;
UCS has several cascading menus. Briefly described in later chapters;
Selection Settings ... in this book the default settings are accepted, and no reference is made to this dialogue box and its settings;
Grips ... will be dealt with later in this chapter;
Drawing Limits page 14. When selected no dialogue box appears; the **LIMITS** prompts appear at the command line.

Fig. 6.1 The **Settings** pull-down menu

The Dimension Style dialogue boxes

The dimensioning of technical drawings is important because many drawings will be of little value until they are fully and correctly dimensioned. AutoCAD 12 dimensioning methods include a complex range of variables which can be set to achieve any form of dimensioning desired – metric, Imperial, ISO, BS 308, architectural, engineering, etc. In this book we are only concerned with explaining how a student or a beginner wishing to learn how to use the software will be able to dimension his/her drawings. As a result, the more complex dimensioning methods will not be found here.

The easiest way to add dimensions to a technical drawing is by using AutoCAD's associative dimensioning features – each dimension with its lines, arrows and figures is treated as a single entity. Each complete dimension entity can be edited with the **Modify** commands such as **ERASE**, **EXTEND**, **ROTATE** and **STRETCH** in a manner similar to that by which any other entity can be modified (edited).

AutoCAD 12 dimensioning features are controlled by a large range of variables. To see the whole range of these, call the command **DIM**, followed by entering a **?**. A text screen appears which can be scrolled by pressing *Return* to see all the dimensioning variables. However, here we will be relying on setting dimensioning variables only from the various dialogue boxes which appear on screen, either by entering the command **ddim** at the keyboard or selecting **Dimension Style ...** from the **Settings** pull-down menu.

The first dialogue box to appear is the **Dimension Styles and Variables** (Fig. 6.2). By selection from the various names in this dialogue box all dimensioning variables referred to in this book can

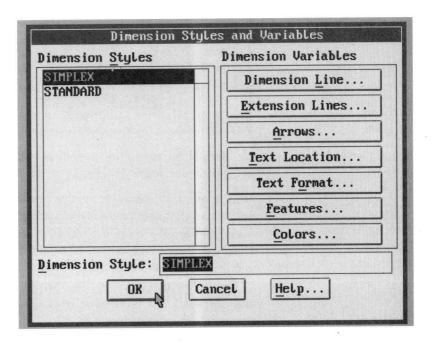

Fig. 6.2 The **Dimension Styles and Variables** dialogue box

be set. As an example select **Extension Lines ...** by moving the arrow cursor on to the words in their box, followed by a *left-click*. The **Extension Lines** dialogue box appears (Fig. 6.3). In this dialogue box it will be seen that:

1. The **Extension Line Color** is to be red – *left-click* on the box by the side of the words. The **Select Color** dialogue box appears (Fig. 4.3 – page 79). Select the box of colour red from this dialogue box;
2. The **Extension Above Line** is to be set at 3 units. Move the arrow cursor on to the numbers box; erase the existing figures by pressing the left arrow key of the keyboard and enter 3;
3. The **Feature Offset** has also been set at 3;

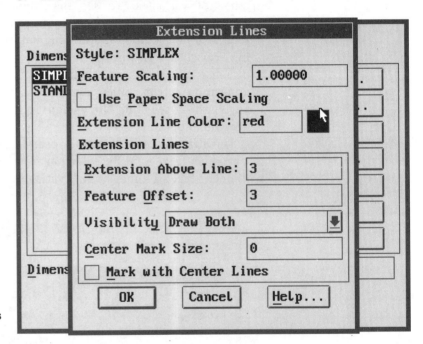

Fig. 6.3 The **Extension Lines** dialogue box

4. Both lines are set to be visible – select **Draw Both**;
5. No **Center Mark** is thought to be required.

In a similar fashion the other variables can be set from their respective dialogue boxes. One further example is given in Fig. 6.4 – all lines are set to appear red on screen by selections made in the **Colors** dialogue box, which relies on the **Select Color** dialogue box referred to earlier.

The following are the settings which will be used here. They can now be included with your *work.dwg* file for use when drawing answers to the exercises:

Dimension Style ... SIMPLEX;
Dimension Line ... red;
Extension Lines ... red
  Extension Above Line 3
  Feature Offset 3
  Center Mark Size 0
  Draw Both
Arrows ...
  Arrow Size 3
Text Location ...
  Text Height 3
  Tolerance Height 1.8 (if needed)
  Horizontal Default
  Vertical Above
  Alignment Align with dimension lines
Features ... shows – Text Style, Extension Lines, Arrows,
                                              Text Position;
Colors ...
  Dimension Line red
  Extension Lines red
  Dimension Text red

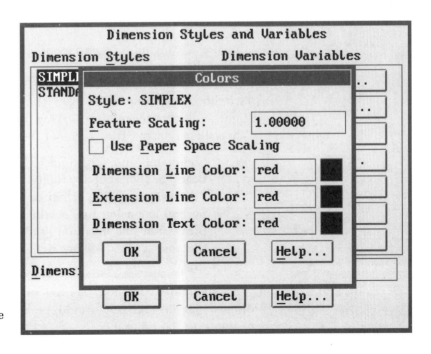

Fig. 6.4 The **Colors** dialogue box

The command DIM

The command can be called from the **Draw** pull-down menu (Fig. 6.5) or by entering dim from the keyboard at the command line. No matter which method is used, the on-screen menu changes to show the various options available within the **Dim** command system. The options we are interested in here are:

Horizontal abbreviation hor
Vertical   abbreviation ver
Aligned   abbreviation ali
Radial   abbreviation rad
Diameter   abbreviation dia
Leader   abbreviation lea

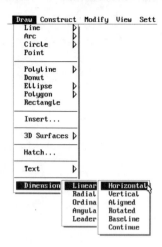

Fig. 6.5 **Dimension** sub-menus in the **Draw** pull-down menu

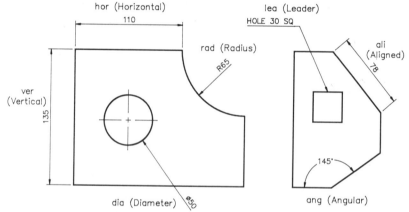

Fig. 6.6 Examples of different types of dimensions

Figure 6.6 includes examples of the dimensions resulting from each of these options.

Taking as an example the option **Horizontal** (abbreviation hor). Two horizontal dimensions are shown in Fig. 6.7, each using slightly different techniques. Taking the upper 140 dimension first. When the command **Dim** is called the command line changes to:

> **Command:** dim   *right-click*
> **Dim:** hor (horizontal)   *right-click*
> **First extension line origin or RETURN to select:**   *pick*
> **Second extension line origin:**   *pick*
> **Dimension line location:**   *pick*
> **Dimension text<140>:**   *right-click* (to accept 140)
> **Dim:**

An alternative method, shown in the lower of the two 140 dimensions in Fig. 6.7, would be to press the *Return* key in response to the prompt:

The Settings menu. Dimensioning 111

First extension line origin or RETURN to select: *right-click* (or Return)
Select line, arc, or circle: *pick*
Dimension line location: *pick*
Dimension text<140>: *right-click* (to accept 140)
Dim:

```
Command: dim  right-click
Dim:       hor (horizontal)  right-click
First extension line origin or RETURN to select:  pick
Second extension line origin:   pick
Dimension line location:        pick
Dimension text <140>:           right-click  (to accept 140)
Dim:
```

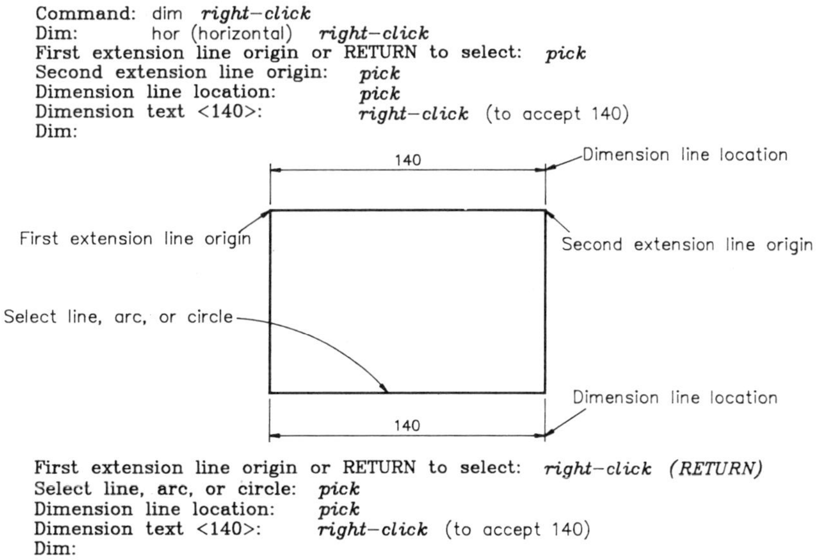

Fig. 6.7 Two examples of horizontal dimensions

```
First extension line origin or RETURN to select:  right-click (RETURN)
Select line, arc, or circle:  pick
Dimension line location:      pick
Dimension text <140>:         right-click  (to accept 140)
Dim:
```

The second method is slightly quicker and can usually be adopted.

If either ver (vertical) or ali (aligned) dimensions are required, the sequence of prompts and responses are similar to those for horizontal dimensions. If, however, arcs or circles are to be dimensioned the following patterns of prompts and responses emerge:

Command: dim *right-click*
Dim: rad *right-click*
Select arc or circle: *pick*
Dimension text<65>: *right-click* (to accept 65)
Enter leader length for text: pick (the text is dragged to a position as the selection device is moved), the *pick* confirms the position when chosen
Dim:

A similar set of prompts and responses occurs for dia (diameter) dimension. Note that the diameter symbol and the letter R for radius are automatically placed in front of the dimension if the final response to **Dimension text<65>:** is accepted. If, however, the operator wishes to enter a different radius or diameter than that given in the < > brackets of the **Dimension text<65>:** prompt, either the letter R must be placed in front of the required new number or, if a diameter, entering **%%c** will produce the required diameter symbol. Similarly, if an angle is being dimensioned and a different angle figure is to be placed in the dimension, entering **%%d** will produce the standard degree symbol behind the number.

To dimension an angle:

> **Command:** dim   *right-click*
> **Dim:** ang (angular)   *right-click*
> **Select arc, circle, line or RETURN:**   *right-click* (or press Return)
> **Angle vertex:**   *pick* (use snap or osnaps to pick the exact vertex position)
> **First angle endpoint:**   *pick*
> **Second angle endpoint:**   *pick*
> **Dimension arc line location (Text/Angle):**   *pick* (dragging the text and arc)
> **Dimension text<145>:**   *right-click* (to accept the 145-degree angle)
> **Enter text location (or RETURN):**   *pick*
> **Dim:**

Tolerances

If dimensions are to show tolerances, they can be set in the **Text Format** dialogue box (Fig. 6.13). An example of tolerances set for a drawing are shown in Fig. 6.8. The settings in the **Text Format** dialogue box are shown with the drawing.

Arrow types

Three different types of "arrow" are included with the standard AutoCAD 12 software. Other types can be easily added by the operator, although it is not the intention to include a description of how this is carried out here. Figure 6.9 shows a simple outline which has had dimensions added showing all three types of "arrow". It should be noted that once the arrow type has been set in the **Arrows** dialogue box, all dimensions within the same drawing will have that type of arrow at the end of each dimension line.

Another example of an outline carrying simple dimensions is given in Fig. 6.10.

The Settings menu. Dimensioning   113

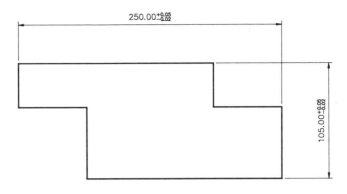

Fig. 6.8 An example of dimensions which include tolerances

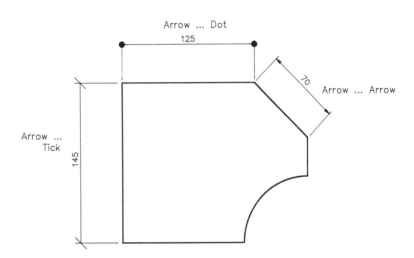

Fig. 6.9 The three types of dimension arrow in AutoCAD 12

The action of Modify commands on dimensions

Figure 6.11 shows how dimensions can be modified with the aid of three of the **Modify** commands (see Chapter 3). The four drawings of Fig. 6.11 show:

1. The original outline with a horizontal dimension – length 120;
2. Both the outline and the dimension have had the **STRETCH** command applied. The resulting outline is stretched, so is the dimension. What is particularly important is that the figures of

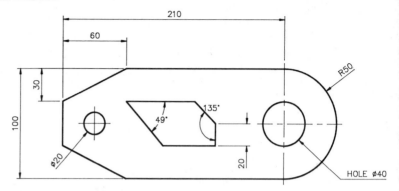

Fig. 6.10 A further example of dimensioning

the dimension have been automatically changed to accommodate the new length of the outline;

3. A vertical line has been added to the drawing and both the outline and the dimension have been **TRIM**med to the vertical line. Again the figure of the dimension has been automatically changed to accommodate the short horizontal length of the outline;

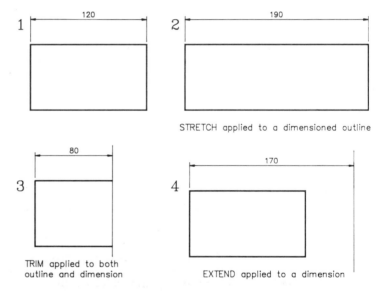

Fig. 6.11 The action of some **Modify** commands on dimensions

4. A vertical line has been placed to the right of the outline and only the dimension has been acted upon by the command **EXTEND**. The resulting dimension adjusts to the new position as to its dimension line, its extension line, its arrow and its figures. This last action can only be used with the linear dimensions – horizontal and vertical.

The Settings menu. Dimensioning

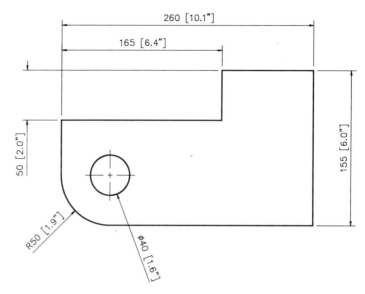

Fig. 6.12 Alternative dimension units

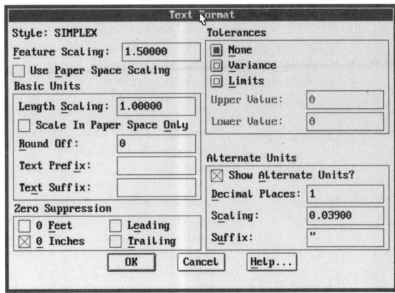

Fig 6.13 The **Text Format** dialogue box showing the **Alternate Units** settings for Fig. 6.12

Alternative units in dimensions

Figure 6.12 is an example of an outline dimensioned with two types of units – millimetres (metric) and inches (Imperial). The settings of the variables to obtain this form of dimensioning are set in the **Text Format** dialogue box. The settings for the dimensions shown in Fig. 6.12 can be seen in Fig. 6.13. The **Show Alternate**

Units box has been selected with a **left-click**; the figure 1 has been entered in the **Decimal Places** box; the **Scaling** is automatically adjusted; the **Suffix** " has been entered into its box. Then going to the **Zero Suppression** area of the dialogue box, the **0 Feet** box is cleared and the **Inches** box is selected. The selection of the **Inches** box, together with the earlier selection of **Decimal** units from the **Units Control** dialogue box, ensures the correct scaling for millimetres to inches in the **Scaling** box of the dialogue box.

Some rules for AutoCAD 12 dimensioning

1. When the positions for extension lines or for the vertices of angles are being *picked*, use **Snap** and **Osnap** to ensure that the points have been accurately selected;
2. Some of the **Modify** commands may be used to edit existing dimensions;
3. The current **Style** of text determines the style of the dimensioning text. Set the text style before adding dimensions;
4. Once **Dim** has been called, dimensions can be added one after the other, without re-calling the command. When one dimension has been added, the command line reverts to **Dim:** ready for the next dimension to be added and not to **Command:** as with other command systems;
5. If a single dimension is to be added to a drawing use the command **Dim1**. This command will not repeat itself after a dimension has been added;
6. Some commands can be called ("transparent" commands) while still in the **Dim** command by selection of the required command from a pull-down menu – e.g. **Zoom** window from the **View** pull-down. Another method is to enter **'zoom**, to obtain the "transparent" command while in **Dim:** – i.e. enter an apostrophe in front of the command name;
7. While in **Dim** the position of the text of a dimension can be moved or edited with the **Dim** command **Tedit**.

Exercises

Note: If you have saved the drawings resulting from answering exercises in earlier chapters, these can now be dimensioned to give practice in dimensioning.

Exercise 1

Figure 6.14. Copy the given plan to the dimensions given. Then fully dimension your drawing.

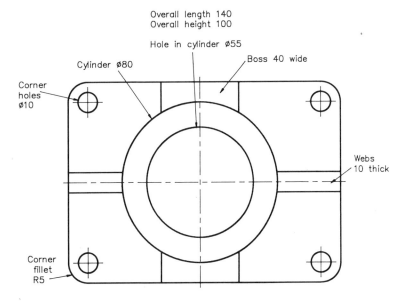

Fig. 6.14 Exercise 1

Exercise 2

Figure 6.15. Draw the given elevation to the sizes given and fully dimension your drawing.

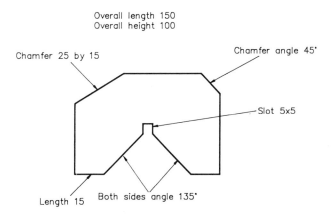

Fig. 6.15 Exercise 2

Exercise 3

Figure 6.16. Construct the hatched interwoven pattern, commencing with a drawing of a pentagon, using the command POLYGON. Hatch and then fully dimension your drawing. Do not copy the grid lines — they are there to show you the size of the edge of the pentagon.

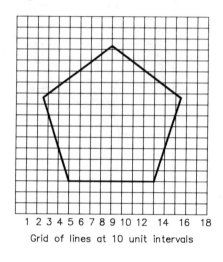

Fig. 6.16 Exercise 3

Grid of lines at 10 unit intervals

Bands are 20 wide
Hatch Pattern is ansi37
Scale 1

Exercise 4

Figure 6.17. Without including the grid network, draw and fully dimension the two patterns.

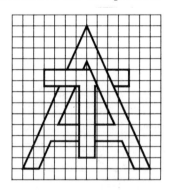
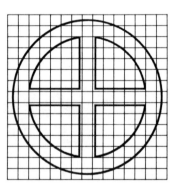

Fig. 6.17 Exercise 4

Grid of lines at 10 unit intervals

Grid of lines at 10 unit intervals

Exercise 5

Figure 6.18. A hatched sectional view of an engineering component is shown. Copy the view to the sizes given. Then add the hatching and fully dimension your drawing.

The Settings menu. Dimensioning 119

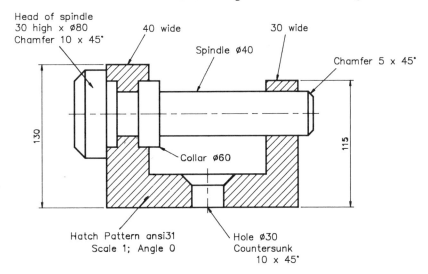

Fig. 6.18 Exercise 5

Grips

To enable **Grips**, select **Grips ...** from the **Settings** pull-down menu. The **Grips** dialogue box appears – Fig. 6.19. Position the cursor arrow over the box at the side of the words **Enable Grips** and *left-click*. **Grips** are enabled if a cross appears within the box. The size of the **Grips** pick-box can also be changed if desired by moving the slider in the **Min, Max** slider bar at the bottom of the dialogue box. The actual size of the pick-box appears in the dialogue box. In Fig.

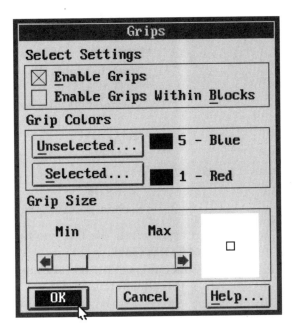

Fig. 6.19 The **Grips** dialogue box

6.19 it will be seen that **Unselected** pick-boxes will be blue and **Selected** pick-boxes will be red. These colours can be changed if other colours are thought to be more suitable.

When **Grips** are enabled, if an entity is *picked* without a command being called, three coloured pick-boxes appear on the selected entity — one at each end of the entity and one at its centre. These pick-boxes are **Unselected** and, if the dialogue box settings are as in Fig. 6.19, will be blue in colour and in outline only. If a second selection is made of any one of the **Unselected** pick-boxes, that pick-box changes colour to red, the outline becomes filled with that colour and is the **Selected** pick-box = the base point of any action required from one of the 5 modifying commands under the control of **Grips**. The use of **Grips** allows an operator to very quickly **STRETCH**, **MOVE**, **ROTATE**, **SCALE** or **MIRROR** those entities to which pick-boxes have become attached. It must be noted that no command is entered or can be called. If the command line shows **Command:** and **Grips** are enabled, when any entity is selected, pick-boxes will appear at its ends and at its centre.

The sequence is as follows:

Command: *pick* an entity (no command has been called).
Grip pick-boxes appear on the entity
Command: *pick* one of the pick-boxes. It changes to a
filled red colour
STRETCH
<Stretch to point>/Base point/Copy/Undo/eXit:
right-click
MOVE
<Move to point>/Base point/Copy/Undo/eXit: *right-click*
ROTATE
<Rotate angle>/Base point/Copy/Undo/Reference/eXit:
right-click
SCALE
<Scale factor>/Base point/Copy/Undo/Reference/eXit:
right-click
MIRROR
<Second point>/Base point/Copy/Undo/eXit:

A *right-click* at this point starts the cycle off again with **STRETCH**.

Examples of the use of **Grips** are given in Fig. 6.20. The drawings of this illustration show:

The Settings menu. Dimensioning   121

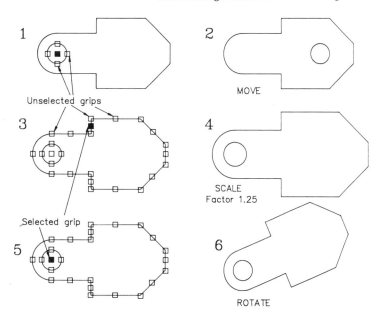

Fig. 6.20 Examples of the uses of **Grips**

Drawing 1

The circle of the drawing was selected and blue pick-boxes appear. The centre of the circle is selected as the **Selected** pick-box.

Drawing 2

The ****MOVE**** option has been chosen – by repeatedly making a *right-click* until ****MOVE**** with its prompts appears at the command line. Then the circle is moved to its new position.

Drawing 3

All entities of the drawing have been selected bringing pick-boxes on all of them. One of these is selected as the base point. That pick-box changes colour.

Drawing 4

The drawing is scaled by a factor of 1.25 after a repeated *right-click* has brought the command ****SCALE**** with its prompts to the command line.

Drawing 5

Again all entities have been chosen and the centre of the circle has been selected as a base point.

Drawing 6

The whole drawing is rotated around the **Selected** pick-box.

The **Grips** dialogue box can be called to screen by entering ddgrips at the keyboard.

CHAPTER 7

Orthographic projection

Introduction

Many of the drawings constructed with the aid of CAD systems will be two-dimensional and the majority of these, particular in industries such as engineering, building, architecture, etc., will be orthographic projections. The theory underlying this form of projection is explained in Appendix A (commencing on page 220). How simple orthographic projections can be constructed with the aid of AutoCAD 12 is the subject matter of this chapter. As will be seen by reference to Appendix A, two forms of orthographic projection are in common use – First Angle and Third Angle. In the past common practice in Great Britain was to use First Angle, while in the USA, Third Angle projection was the norm. However in recent years the use of Third Angle projections is becoming increasingly common in Great Britain. The reader is advised to familiarize him/her self with both angles of projection.

Drawing sheet layout

The draughting/design offices of all industrial concerns will have their own forms of drawing layout sheets designed to meet the company's own requirements. Most colleges will also usually have their own drawing sheet layouts on which students will be expected to construct their drawings. A very simple drawing sheet layout was suggested on page 43. Figure 7.1 is a second sheet layout which includes more spaces for the addition of details connected with the object being described on the drawing sheet. We will be using this sheet layout in later examples in this chapter. If the reader finds the layout shown in Fig. 7.1 to be suited to his/her requirements, copy the given layout and replace the drawing file *work.dwg* with the new layout. Note that the configuration details applying to the original *work.dwg* file can be used for this new file. The configuration details are:

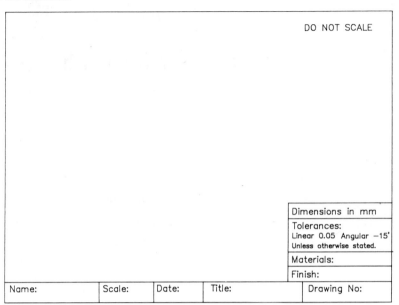

Fig. 7.1 An example of a drawing sheet layout

1. Limits set to 0,0 and 420,297 (A3 sheet size in millimetres);
2. **Grid** set at 10 and on; **Snap** set at 5 and on; remember **Grid** can be toggled on or off with **F7**; **Snap** can be toggled on/off with **F9**;
3. **Style** for text set for **Simplex** of height 6;
4. **Dim** variables:
   Text Style Simplex;
   Dimension Lines red;
   Extension Lines 3 unit extensions on both; colour red;
   Text height 3; **In Line** with dimension lines; **Above** dimension lines; colour red;
   Arrows length 3; **Arrows**;
5. Layers as follows:
   0 standard layer;
   Construction colour magenta; continuous lines;
   Centre colour blue; centre lines;
   Hidden colour red; hidden lines;
   Text colour cyan; continuous lines;
   Dimension colour green; continuous lines;
6. When adding text in the Name, Scale, Date, Title and Drawing No. boxes change the text Height to 8.

Constructing a First Angle orthographic projection

The series of drawings in Figs 7.2 to 7.6 illustrate a sequence of

Orthographic projection 125

work suitable for constructing the details for a simple coupling component in either First or Third Angle orthographic projections. Although Figs 7.2 and 7.3, as well as the later Fig. 7.7, do not include the sheet layout given in Fig. 7.1, the reader can, if he/she thinks fit, start by opening the drawing sheet file *work.dwg* and constructing the projection within the boundaries of the work layout sheet. The component being drawn is illustrated by the two pictorial views of drawing 1 of Fig. 7.2. The sequence for the construction of the three views follows a pattern such as:

1. Open the *work.dwg* file;
2. Make sure **Grid** is on (F7); make sure **Snap** is on (F9);
3. Set the layer **Construction** as the current layer;

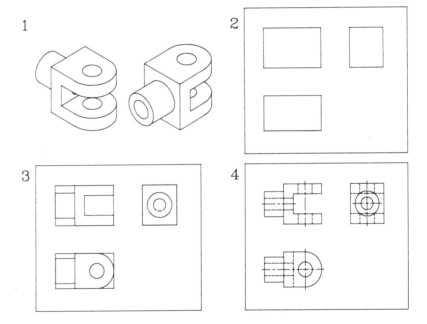

Fig. 7.2 Stages in constructing a First Angle orthographic projection

4. Drawing 2 of Fig. 7.2. Draw rectangles of overall width and overall height for the three views with the aid of the command **LINE**. Use snap points and, if necesary, **Osnaps** to ensure the outlines are accurate as to dimensional sizes. Construct the rectangles with the aid of the **Relative coordinate** method (page 26);
5. Although an attempt should be made to achieve good positioning of the views to attain a good layout on the drawing sheet, remember that the command **MOVE** can be employed at any time to re-arrange views to better positions in relation to each other;

6. Drawing 3 of Fig. 7.2. Set Layer **0** as the current layer. Using **Snap** and, if necessary, **Osnaps**, construct accurate outlines of the three views, basing their dimensions on the previously constructed lines on the **Construction** layer;
7. Drawing 4 of Fig. 7.2. Turn layer **Construction** off. Make the layer **Centre** current and add all necessary centre lines. Make the layer **Hidden** current and add all hidden detail lines;
8. Figure 7.3. Set layer **Dimensions** and add all dimensions;
9. Figure 7.4. If working within a sheet layout, set layer **text** as current and add the statement **First Angle projection** and fill in details, such as your Name, the Scale, etc., within the various boxes;
10. Save the drawing to a suitable filename – e.g. **COUPLING**. The extension .dwg will be automatically added by AutoCAD 12.

Notes

1. If working on a complicated projection, it is advisable to save the drawing to a named file at intervals – say every fifteen minutes. If an unexpected power cut or some other form of interference prevents the drawing from being fully completed, at least that part up to the last **Save as** will be on file and can thus be called back to the screen for further work;
2. Figure 7.3 shows the views of the projection drawn with the aid of the command **LINE**. Figure 7.4 shows the same views drawn with the aid of the command **PLINE**. In this example the

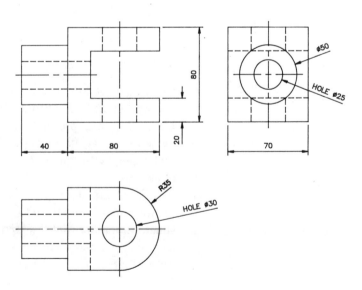

Fig. 7.3 The completed projection

Orthographic projection 127

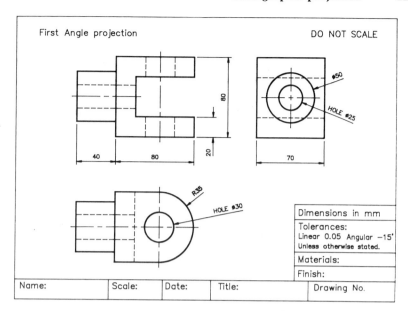

Fig. 7.4 The same projection with polyline outlines and drawn in a drawing sheet

```
Command:      pl (pline) right-click
From point:   pick
Arc/Close/Halfwidth/Length/Undo/Width/<Endpoint of arc>: w (Width) right-click
Starting width <0.0>: 0.7 right-click
Ending width <0.7>:   right-click  (to accept 0.7)
Arc/Close/Halfwidth/Length/Undo/Width/<Endpoint of arc>:   a (Arc) right-click
Angle/CEnter/CLose/Direction/Halfwidth/Line/Radius/ Second pt/
              Undo/Width/<Endpoint of arc>: ce (CEnter) right-click
Center point:              pick
Angle/Length/<Endpoint>:   pick
<Endpoint of arc>:         pick
```

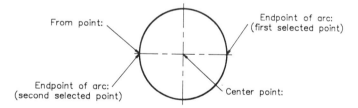

Fig. 7.5 Drawing a circle with the command **PLINE**

polyline thickness was set at 0.7. The advantage of drawing views with polylines is that the outlines stand clear from hidden lines, centre lines and dimension lines and thus make the drawing easier to read;

3. When drawing the arcs and circles of a view in orthographic projection with the command **PLINE**, follow the procedure given in Fig. 7.5.

Constructing a Third Angle orthographic projection

A Third Angle orthographic projection can be constructed in the same manner as a First Angle projection, the only real difference being the positioning of views in relation to each other. Another method would be to draw the required views in First Angle and then, with the aid of the command **MOVE**, re-position the views from First to Third Angle. This is how the projection in Fig. 7.6 was constructed.

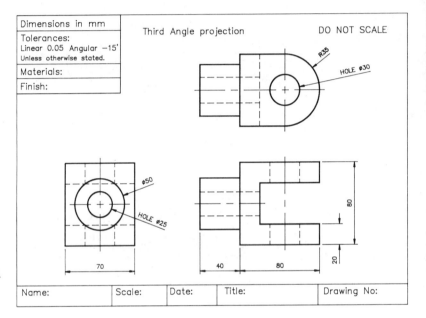

Fig. 7.6 The projection changed to Third Angle with the aid of the **MOVE** command

Sectional views

Figure 7.7 shows a sectional front view and a plan of the coupling in both First and Third Angles. Figure 7.8 is a pictorial view of the coupling showing how the theoretical method of how the sectional view was obtained – the front half of the coupling was cut away and the cut surface hatched with lines at 45 degrees and 4 units apart.

Exercises

In order to practise the construction of orthographic projections, a number of exercises for both First and Third Angle are given below. These exercises are arranged in increasing difficulty from very easy

Plate I The **Hatch Options** dialogue box selected from the **Boundary Hatch** dialogue box, together with **Hatch Patterns** selected from **Hatch Options**. Colours of dialogue boxes set with the aid of the command **DLGCOLOR**.

Plate II The **Select Text Style** dialogue box from **Entity Creation Modes**, together with the **Text Style ROMANC Symbol Set** dialogue box from **Select Text Style**. Colours of dialogue boxes set with the aid of the command **DLGCOLOR**.

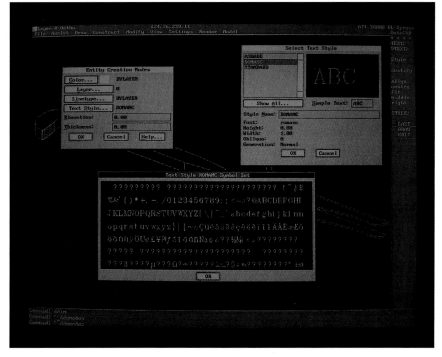

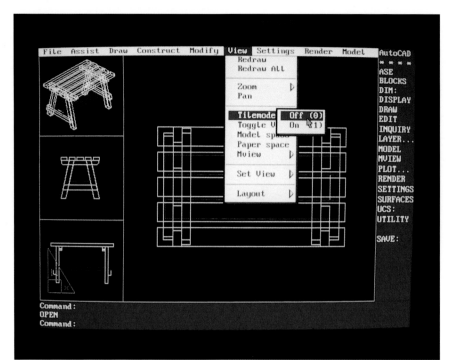

Plate III A 4-viewport screen in **Paper Space** with the pull-down menu **View** showing that **Tilemode** has been set **Off**. The AME solid model was constructed in **Model Space.**

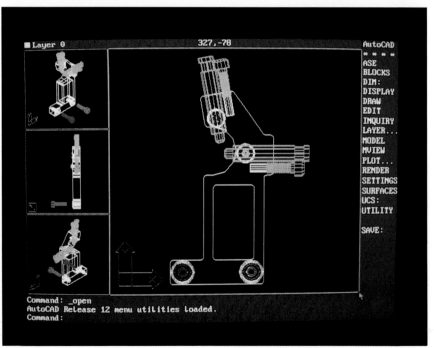

Plate IV A 4-viewport screen in **Model Space** showing four views of an AME solid model.

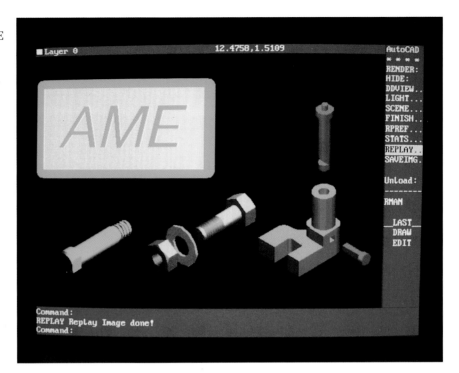

Plate V A 256-colour rendering of several AME solid models. Rendering with the AutoCAD Release 12 **AVE Render.**

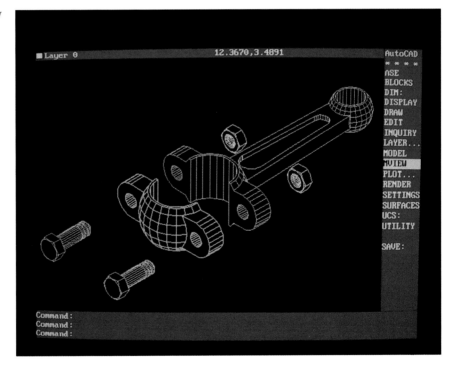

Plate VI A **Vpoint** view of an exploded solid model – a crank from a small compressor – constructed with the aid of AME after **HIDE** has removed hidden lines.

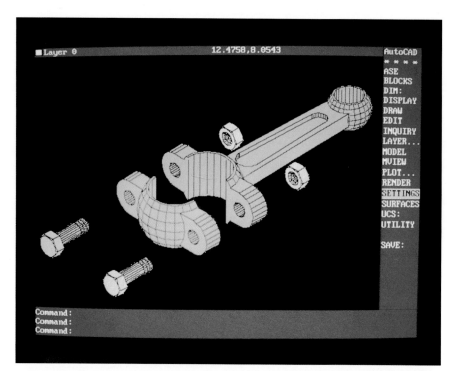

Plate VII The AME solid model of Plate VI after the action of the command **SHADE**.

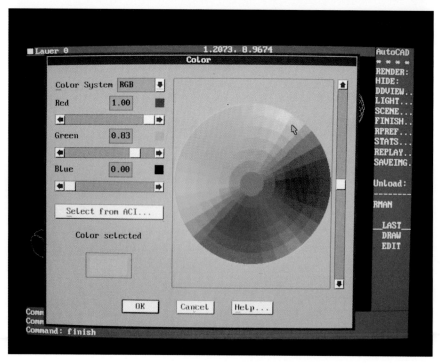

Plate VIII Selecting a colour for one part of the solid model (Plate VI) from the colour wheel in the **Color** dialogue box.

Plate IX Checking lighting effects for the colour selected from the colour wheel (Plate VIII) in the **Modify Finish** Render dialogue box.

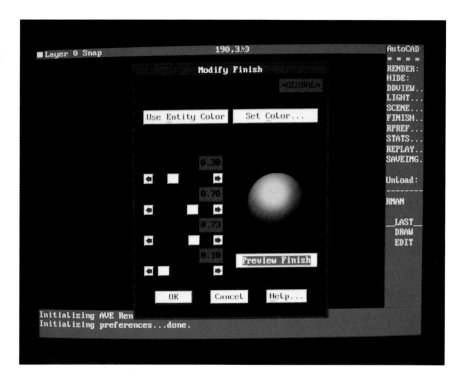

Plate X The fully rendered exploded solid model of Plate VI. Rendering in AutoCAD Release 12 with **AVE Render.**

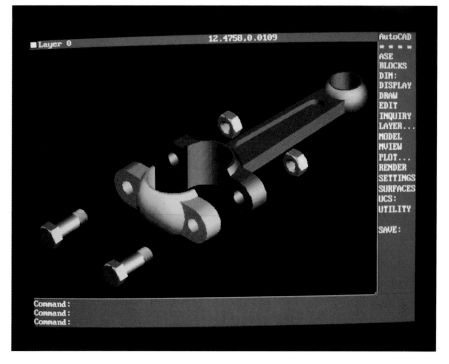

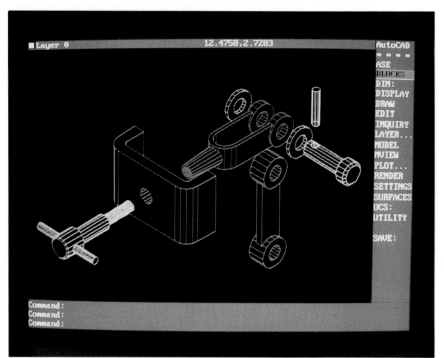

Plate XI A design for a coupling linkage. AME solid models.

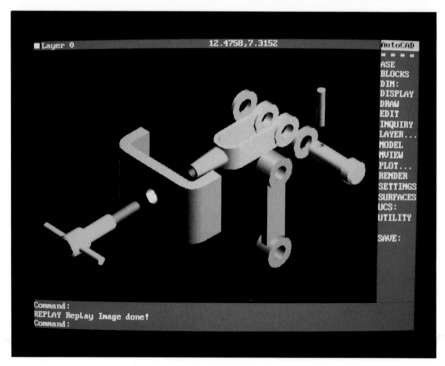

Plate XII The design of Plate XI rendered with **AVE Render.**

Plate XIII An AME solid model of a bungalow.

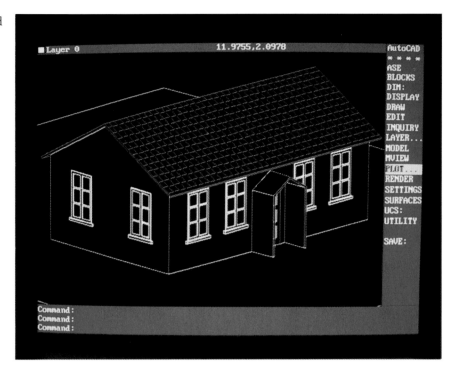

Plate XIV The bungalow of Plate XIII in a different **Vpoint** position after rendering with the aid of **AVE Render**.

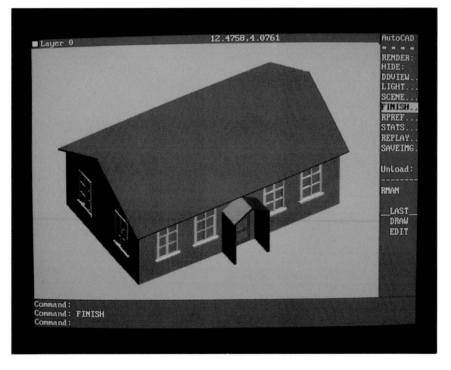

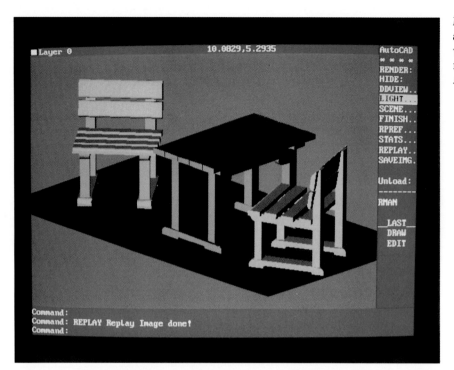

Plate XV A garden table and chairs constructed with the aid of AME and rendered with the aid of **AVE Render.**

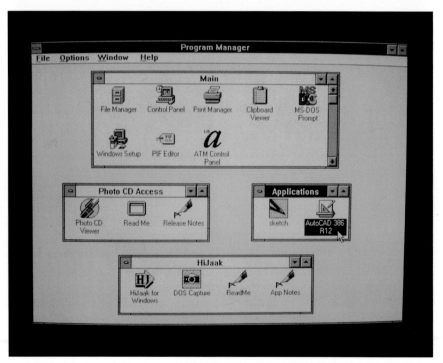

Plate XVI AutoCAD 386 Release 12 can be run as an **MS-DOS** window in Windows 3.1.

Orthographic projection   129

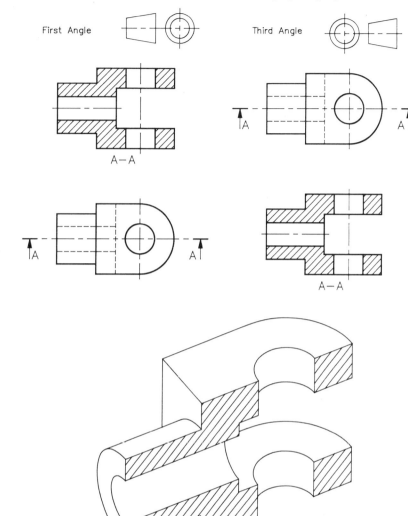

Fig. 7.7 The plan and front sectional view compared in both First and Third Angle projections

Fig. 7.8 A pictorial view showing the theoretical idea of a sectional view

to more difficult. Help may be had by turning to the pages dealing with the underlying theory of orthographic projection in Appendix A (pages 220 to 226).

Exercise 1

A support stand is shown by a pictorial view and a plan in Fig. 7.9. Copy the given plan and, working to the information given in Fig. 7.9, add the following views in either First or Third Angle orthographic projection:

(a) a sectional front view with the section plane passing through the horizontally placed ribs in the plan;
(b) an end view.

Completely dimension your three views.

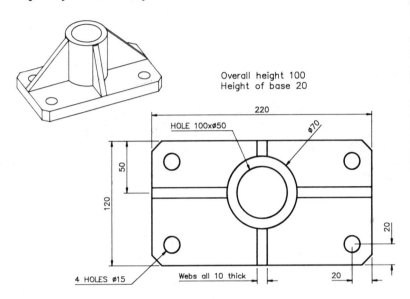

Fig. 7.9 Exercise 1

Exercise 2

Figure 7.10 shows a pictorial view together with a Third Angle orthographic projection of a plan and a front view of part of a grinding machine tool holder. The pin for positioning the tool jig is shown in the pictorial view, but not in the plan and front view.

Construct in either First or Third Angle projection:

(a) the given plan;
(b) a sectional front view with the pin fully positioned within its holes and with the section plane passing centrally along the pin;
(c) an end view.

Fully dimension your three views.

Exercise 3

Figure 7.11 is a pictorial view of a picnic area table.

Working to the dimensions given and using your discretion about dimensions not included with the pictorial view, draw the

Orthographic projection   131

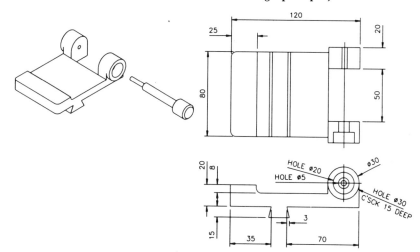

Fig. 7.10 Exercise 2

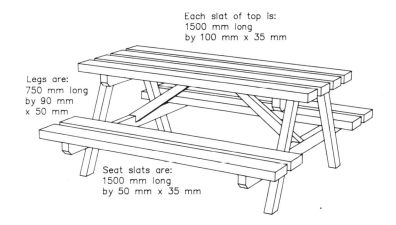

Fig. 7.11 Exercise 3

following three views in either First or Third Angle orthographic projection:

(a) a front view;
(b) an end view;
(c) a plan.

Fully dimension your drawing.

Exercise 4

Figure 7.12 is a pictorial drawing of a garden shed. All windows and doors are shown – the two walls not shown do not have either windows or doors. Working to a scale of 1:20 to the dimensions included with Fig. 7.12 and using your own discretion about sizes and dimensions not shown, construct the following orthographic views in either First or Third Angle projection:

(a) taking the side of the shed with four windows as the front, draw a front view;
(b) taking the side with the door as the end, draw an end view;
(c) a plan.

Fully dimension your drawing.

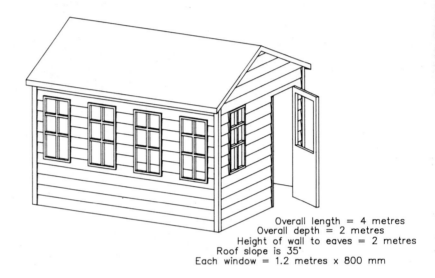

Overall length = 4 metres
Overall depth = 2 metres
Height of wall to eaves = 2 metres
Roof slope is 35°
Each window = 1.2 metres x 800 mm

Fig. 7.12 Exercise 4

Exercise 5

A car jack is shown in Fig. 7.13. In the pictorial view, the screw of the jack has been removed to show how the pins on the screw-supporting spindles are intended to fit into the moving parts of the jack. Working to the dimensions included with the drawings and working in either First or Third Angle orthographic projection draw the following views of the jack:

(a) a front view;
(b) a sectional end view with the section plane passing through either one of the screw-supporting spindles;
(c) a plan.

Use your discretion about sizes not included in Fig. 7.13.
Fully dimension your views.

Orthographic projection

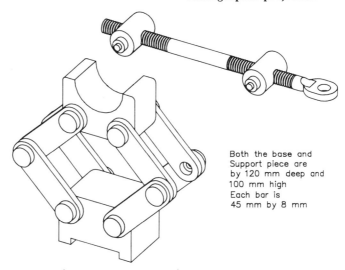

Both the base and Support piece are by 120 mm deep and 100 mm high
Each bar is 45 mm by 8 mm

Fig. 7.13 Exercise 5

Exercise 6

Figure 7.14 shows a pictorial view and a plan of a spindle support with a cap for covering one end. To the dimensions given and working in either First or Third Angle orthographic projection, construct the following:

(a) a front view looking at the front of the pictorial view;
(b) a sectional end view with the cap in position at one end and with the line of the section plane passing through the central web;
(c) a plan.

Fully dimension your views.

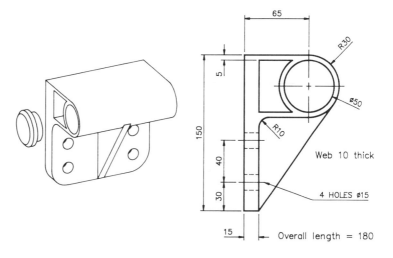

Fig. 7.14 Exercise 6

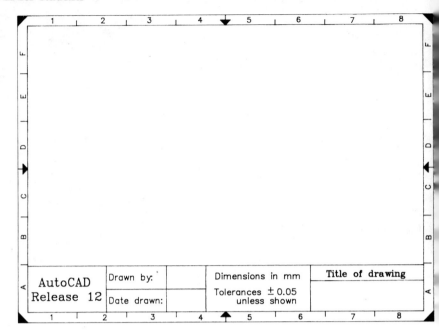

Fig. 7.15 Another example of a drawing sheet layout

CHAPTER 8

Wblocks, blocks and inserts

Introduction

Any AutoCAD drawing file can be included in the AutoCAD drawing currently on screen with the aid of the command **INSERT**. This allows items such as symbols, drawings of small components and any other drawing held on file to be added to the current drawing. This facility is of particular value when constructing circuit drawings or drawings which rely heavily upon symbols. It also allows the building of "libraries" of files on disk of drawing symbols for inclusion in any drawing. The use of such libraries enables much speedier construction of many drawings and also reinforces the knowledge that with CAD an operator need *never draw the same thing twice*. Libraries of such features as the following can either be purchased from specialist software suppliers, or built up as the operator wishes:

1. Electrical and electronics symbols for circuit drawing;
2. Pneumatic and other fluidic symbols for constructing pneumatic and hydraulic circuits;
3. Gate logic symbols for logic circuit drawing;
4. Building and architectural drawing symbols for use in architecture;
5. Engineering fastening symbols such as bolts, nuts, studs, rivets, etc., for inclusion in engineering drawings;
6. Dimensioning symbols such as geometrical tolerances for inclusion with the dimensioning of views;
7. Sheet size drawings, complete with title block and other features, ready to be added to projections.

The command WBLOCK

The method of compiling a library of symbols involves using the command **WBLOCK**. Figure 8.1 shows the sequence of commands,

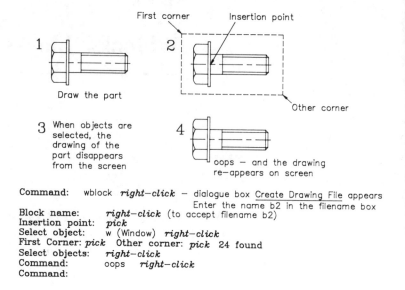

Fig. 8.1 The procedure for creating a **WBLOCK**

prompts and options for constructing a symbol of a bolt and washer with this command. When the command sequence is completed, a standard AutoCAD drawing file, with the extension .*dwg* will have been saved to disk. In the example in Fig. 8.1, the filename B2 has been entered to be saved to disk B:\. The file of the bolt and washer will be saved to disk with a filename of *b:\b2.dwg*. A **WBLOCK** is a standard AutoCAD drawing file. The command sequence is:

Command: wblock   *right-click*
  the **Create Drawing File** dialogue box automatically appears (Fig. 8.2). Enter the required filename (B2) in the box to the right of **File:**, followed by a *left-click* on **OK**
Block name: *right-click* (to accept the name B2)
Insertion point: pick a suitable point on the drawing outline
Select object: w (window)   *right-click*
First corner: pick   **Other corner:**   pick **24 found**
Select objects: *right-click* (to confirm selection within window)
  the entities selected within the window disappear from the screen. They have been saved with the filename
    b:\b2 to disk **B:**
Command:

If, at this stage, the entities saved to file are required in the

Wblocks, blocks and inserts 137

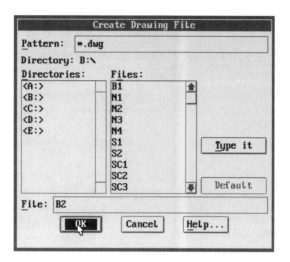

Fig. 8.2 The **Create Drawing File** dialogue box

current drawing, enter the command **oops** at the command line and the entities return to the screen. Thus:

    Command: oops *right-click*
    Command:

Note: The insertion point must be chosen with care. Blocks must usually be inserted in exact positions in the current drawing on screen. Use **Snap** or **Osnap** points to ensure accuracy of selection of the insertion point.

The command DDINSERT

Any drawing (remember a **WBLOCK** is a drawing) can be inserted into the current drawing on screen. There are two methods. The first is to enter the command **INSERT** at the command line. The second calls up the dialogue box **Insert** by entering the command **DDINSERT** at the command line. When the second of these two commands is chosen, the command line shows:

    Command: ddinsert *right-click*
        The **Insert** dialogue box appears (Fig. 8.3). The **Insert** dialogue box allows either the filename to be entered in the **File ...** box, or a *left-click* on **File ...** brings the **Select Drawing File** dialogue box on screen over the **Insert** dialogue box (Fig. 8.4). From this new dialogue box, first *left-click* on the name of the disk holding the required

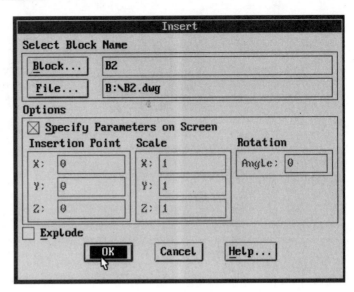

Fig. 8.3 The **Inserts** dialogue box

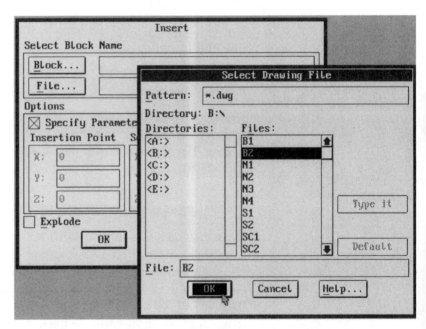

Fig. 8.4 The **Select Drawing File** dialogue box from the **Inserts** dialogue box

file, then *left-click* on the filename in the list appearing under **Files:**. If this is followed by a *left-click* on **OK**, and a second *left-click* on **OK** in the **Insert** dialogue box, the wblock drawing appears on screen attached to the cursor by its **Insertion point**. The command line changes to:
Block name (or ?): *right-click* (this accepts the filename)

Wblocks, blocks and inserts 139

Insertion point: *pick*   **<X scale factor>/Corner/XYZ:**
right-click (to accept the drawing at its original X scale)
Y scale factor (default=X):   *right-click* (to accept the
drawing at its original Y scale)
Rotation angle <0>:   *right-click* (to accept the original
drawing angle)
Command: and the wblock drawing appears on screen
full size at angle 0.

Notes

1. If the variable **DRAGMODE** is set to **Auto**, the wblock drawing appears at the cursor and can be dragged across the screen to be inserted at any required point;
2. The inserted drawing can be scaled to any size (Fig. 8.5) – a number greater than 1 enlarges the drawing, a number less than 1 decreases the scale of the figure. If a negative number is entered, the drawing is mirrored along the axis to which the negative number is applied;
3. The number given in response to the **Rotation angle** should be in degrees (unless AutoCAD is configured otherwise) and the usual angular measurement is taken as being anti-clockwise with 0 (360) degrees being horizontal East (Fig. 8.5);
4. An inserted drawing can be treated as a single entity, unless

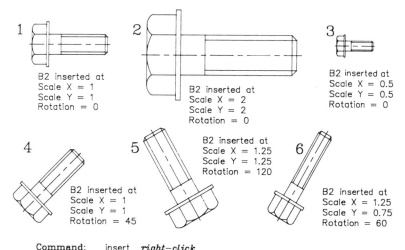

Fig. 8.5 Inserting a block to different scales and rotation angles

the **Explode** box of the **Insert** dialogue box is checked, when the drawing explodes into its original entities;

5. Once the drawing is on screen it can be exploded by entering the command **EXPLODE**, followed by selecting any point on the wblock drawing. The drawing then explodes into its original drawing entities;
6. If a drawing without a previously selected **Insertion point** is inserted, the **Insertion point** is assumed to be x,y=0,0;
7. The **XYZ** of the option seen above shows that an **Insertion point** can be picked in 3D;
8. Although an insert can be made without calling the **Insert** dialogue box to screen, the advantage of using the dialogue box is that the operator can pick from named files in the **Select Drawing File** box. If the command **INSERT** is used, the filename and the disk holding the file must be known to the operator;
9. Wblocks should be inserted only on **Layer 0**. If inserted on any other layer, the drawing will assume the colour and linetype of that layer. If, for example, the bolt (file B2) has been inserted on layer **Hidden**, the whole drawing would appear on screen in hidden lines;
10. Once the wblock is inserted and exploded, it can be acted upon by the command **STRETCH** (Fig. 8.6).

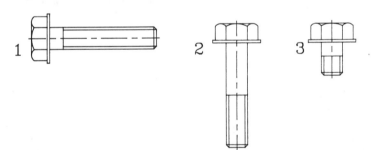

Fig. 8.6 The effect of the command **STRETCH** on blocks

Blocks

When a number of wblocks have been inserted into a drawing, their filenames can be seen by a *left-click* on the name **Block ...** in the **Insert** dialogue box. The **Blocks Defined in this Drawing** dialogue box then appears – Fig. 8.7.

If a ? is the response to the **Block name (or ?):** prompt of the **Insert** command sequence, the drawing editor screen flips to a text screen and a list showing the number and type of blocks in the drawing appears on screen.

Wblocks, blocks and inserts

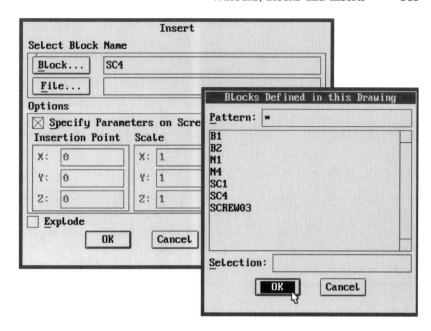

Fig. 8.7 The **Blocks Defined in this Drawing** dialogue box from the **Inserts** dialogue box

Any block held in a drawing can be re-installed in the drawing a second or more time(s) by a *left-click* on the drawing name in the **Blocks Defined in this Drawing** dialogue box.

Libraries of blocks

As mentioned earlier (page 135), libraries of blocks can be compiled by the operator or disks of libraries can be purchased from specialist software suppliers. Some purchased libraries may

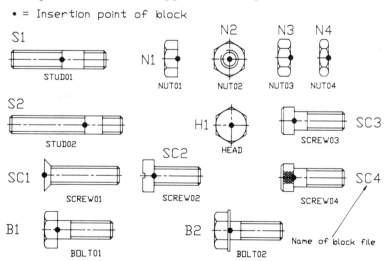

Fig. 8.8 A library of symbols of engineering parts

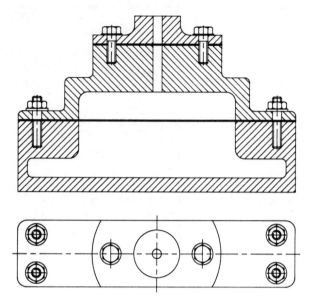

Fig. 8.9 An example of symbols inserted into a drawing from those in Fig. 8.8

contain thousands of symbol block drawings. The insertion of blocks from previously constructed drawings is a great time saver when drawing with the aid of AutoCAD. Here, in this book, only examples of small libraries of blocks which can be constructed by the reader are shown. The first of these – Fig. 8.8 – is of a group of bolts, nuts and washers such as may be inserted in an engineering drawing. Figure 8.9 is an example of a sectional view which includes some of the blocks from Fig. 8.8, inserted to a scale of 0.5 and rotated through either 90 or 270 degrees.

The construction of circuit drawings from blocks

Figure 8.10 shows a small library of electrical and electronics symbols such as could be constructed by the reader. Figure 8.11 is an example of a simple circuit diagram built up from this library. It must be noted that the sequence given in this example is not necesarily that which might be adopted by the operator. When constructing such circuit diagrams all operators will develop their own methods. The given example follows the sequence:

Drawing 1

Insert the symbol blocks at approximate positions in the drawing screen. Make full use of snap points, ensuring that the insertion points of the blocks are at snap points.

Wblocks, blocks and inserts

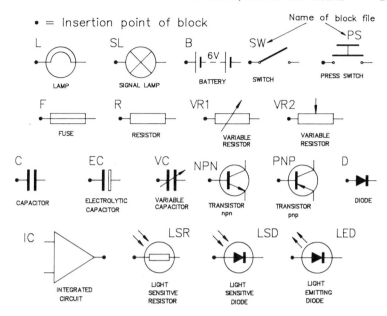

Fig. 8.10 A library of electrical and electronics components

Drawing 2

Move the inserts to more exact positions on the screen, again making full use of snap points.

Drawing 3

Add the conductor lines. Place donuts at conductor junction points. Explode the bottom right-hand resistor, erase its bottom line

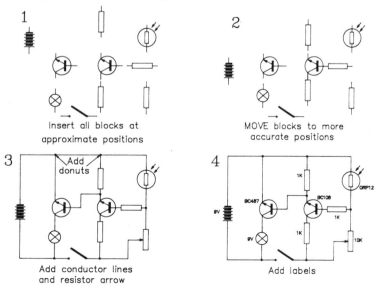

Fig. 8.11 The sequence of constructing a simple electronics circuit from blocks

and add the arrow and lines to show it is a variable resistor of the type required in the circuit. Move blocks and/or lines as necessary to achieve a good circuit layout.

Drawing 4

Add labels to show the names and sizes of the components.

Figure 8.12 is another example of a circuit drawing with symbols inserted from the blocks of Fig. 8.10.

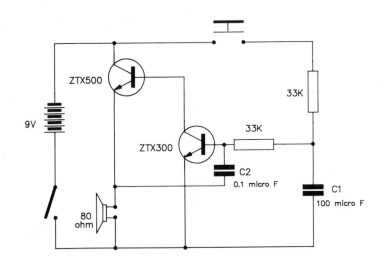

Fig. 8.12 A second example of an electronics circuit constructed by inserts from Fig. 8.10

Other examples of libraries

Pneumatics circuit symbols

A small library of pneumatics circuit symbols is given in Fig. 8.13 and a circuit derived from this library is shown in Fig. 8.14. The circuit shows the depth of the drilling of a hole under the control of a pneumatics circuit.

Logic gate symbols

A library of the main logic gate symbols is shown in Fig. 8.15; Fig. 8.16 shows a circuit derived from these symbols.

Wblocks, blocks and inserts 145

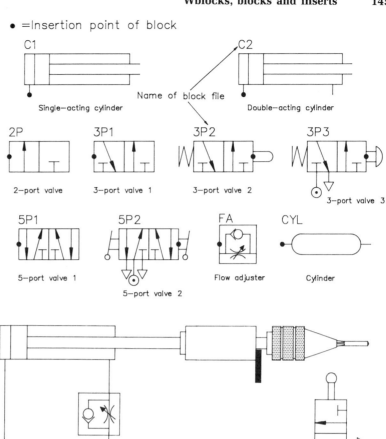

Fig. 8.13 A small library of pneumatics components symbols

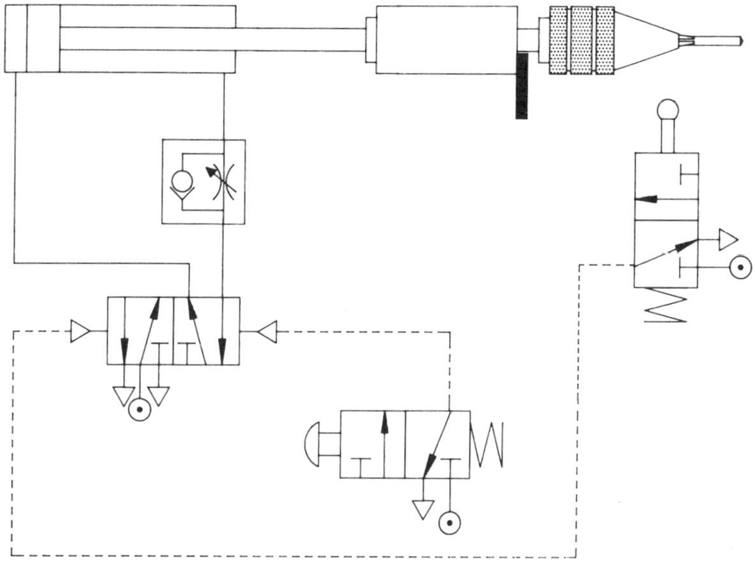

Fig. 8.14 A simple pneumatics circuit constructed from symbols from Fig. 8.13

Building drawing symbols

Figure 8.17 contains some building drawing symbols and Fig. 8.18 a plan of a bungalow showing the outline of its rooms constructed from symbols in the library.

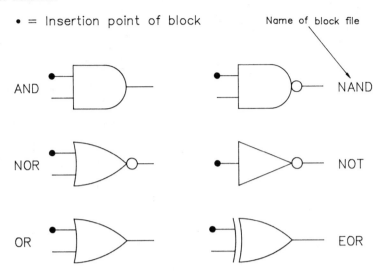

Fig. 8.15 A library of gate symbols

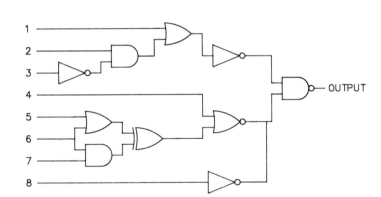

Fig. 8.16 A gates circuit constructed from inserts from Fig. 8.15

Exercises

Exercise 1

Copy sufficient of the symbols from Fig. 8.10 to construct the circuit of Fig. 8.12. Save each of the symbol drawings as a **WBLOCK**. Then, by insertion of the blocks you have saved, construct the circuit diagram Fig. 8.12.

Wblocks, blocks and inserts 147

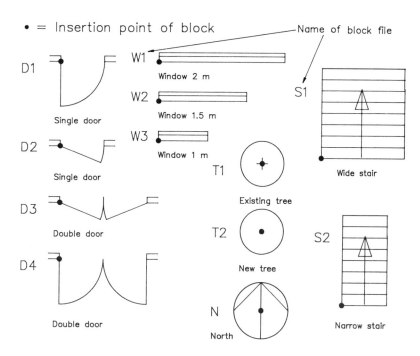

Fig 8.17 A library of building drawing symbols

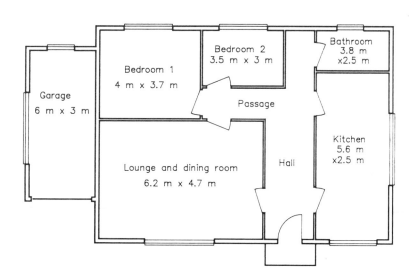

Fig 8.18 A plan of a bungalow layout using some of the symbols from Fig. 8.17

Exercise 2

In a similar manner, copy symbols from Fig. 8.13 to construct the circuit diagram Fig. 8.14.

Exercise 3

Using the same method as for Exercises 1 and 2, copy symbols from Fig. 8.15 and then construct the circuit diagram Fig. 8.16.

Exercise 4

Repeat this procedure – taking symbols from Fig. 8.17 – to construct the bungalow drawing Fig. 8.18.

CHAPTER 9

Pictorial drawing

Introduction

In this chapter pictorial drawing in isometric, cabinet and planometric forms will be briefly described. The more important 3D pictorial methods will be described in later chapters. Pictorial drawings in isometric, with two axes at 30 degrees and a third at 90 degrees can be readily constructed in AutoCAD. Drawings constructed in isometric are not three-dimensional, although they appear, at first sight, to be so. Only the two coordinate axes x and y are employed. The third 3D axis z is not brought into isometric drawing. Isometric drawing entails several command systems – **Snap**, **Isoplane** and **Ellipse**. The two commands which rely on the coordinate system – **Coords** and **Ortho** – can also play an important part in this form of drawing.

The command SNAP

It will be remembered that when **SNAP** is set **ON**, if one *picks* any point on the screen, the cursor locks to the nearest snap point. Selection of points between snap points results in the cursor locking at the nearest snap point. Enter **snap** at the command line and the command line changes to:

> **Command:** snap *right-click*
> **Snap spacing or ON/OFF/Aspect/Rotate/Style<5.00>:**

These options give the following results when chosen:

Spacing: sets the unit distance between snap points on the screen in both x and y directions. A *right-click* in response to the above set of options results in the snap spaces being set to 5 units in both directions on the screen;

ON: sets snap on, so that *picked* points lock at snap points;

OFF: switches snap off, allowing any point on the screen to be selected;

Aspect: entering an **a** allows horizontal spacing between snap points to be different to the vertical spacing. The snap points can thus be set to a rectangular pattern, rather than the more common square pattern;

Rotate: allows the snap point settings to be set at an angle to the x axis. Enter **r** in response to the snap options and the prompts change to **Base point<0.00,0.00>:**. Although a new base point can be entered, the usual practice when rotating the snaps is to accept the 0,0 default. When accepted by a *right-click* entering an angle will cause all the snap points to rotate to the given angle, although they remain in their square or rectangular pattern. If **GRID** is on the grid, points change with the snap rotation;

Style: snap has two styles – **S**tandard and **I**sometric. If the response to the snap options is an **s** (Style) the prompts change to **Standard/Isometric <5.00>:**. Entering **i** (Isometric) changes the pattern of the snap points to one which lies at 30 degrees to both left and right. The setting figure (in this case 5) is the snap spacing along the y axis. The complete set of options and responses to obtain an isometric snap pattern is:

> **Command:** snap   *right-click*
> **Snap spacing or ON/OFF/Aspect/Rotate/Style<0.00>:** s
>
>   (style)   *right-click*
> **Standard/Isometric<5.00>:** i (isometric)   *right-click*
> **Vertical spacing<5.00>:**   *right-click* (to accept the spacing
>   of 5 units vertically – i.e. along the y axis)
> **Command:**

and, if **GRID** is on, the grid pattern changes to an isometric pattern.

The command ISOPLANE

When constructing an isometric drawing the three faces – top, left and right of the drawing – will have axes aligned in different directions. To allow the alignment to be chosen, call **ISOPLANE**. This results in the following at the command line:

> **Command:** isoplane   *right-click*
> **Left/Top/Right<Toggle>:**

The response to this requires pressing the *Return* key (not a *left-click*), to accept the current isoplane. Repeated pressing of the

Pictorial drawing 151

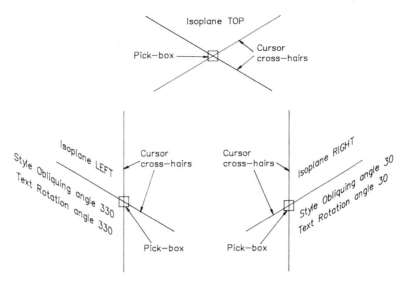

Fig. 9.1 The three isoplanes of isometric drawing in AutoCAD

Return key will toggle between the three in the order shown in the options – left/top/right.

However, it is easier to toggle between the isoplanes by pressing **Ctrl+E** keys. Repeated pressing of these two keys toggles between the isoplanes in the same order. As the isoplane changes in response to the toggling, the cursor hairs move to new positions – Fig. 9.1 shows the movements of the cursor hairs in response to each isoplane position.

The command ELLIPSE

If the isometric snap grid is on ellipses in isometric can be drawn. If snap is not set for isometric then it is not possible to draw isometric ellipses – the isometric ellipse option just does not appear when the command ellipse is called. To draw an ellipse in an isometric drawing:

Command: ellipse   *right-click*
<Axis endpoint 1>/Center/Isocircle: i (isocircle)   *right-click*
Center of circle: *pick or key coordinates*
<Circle radius>/Diameter: *pick or enter a number*
Command:

If further isometric circles (ellipses) are to be drawn, each one requires the i (isocircle) response to the **Ellipse** command.

Note: An isometric ellipse will be drawn on the current isoplane, although this can be altered by toggling after asking for an isocircle

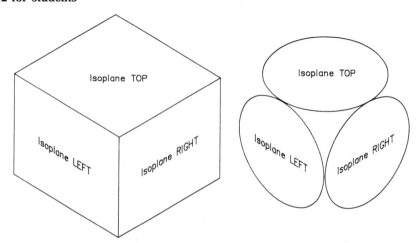

Fig. 9.2 Lines and isocircles drawn in the three isoplanes

and before the radius/diameter of the ellipse is entered.

Figure 9.2 shows the following:

1. Three faces of a cube drawn in isometric with the name of the isoplane on each face;
2. Isometric circles (ellipses) drawn on each of the three isoplanes;
3. Text showing the isoplanes. Referring back to Fig. 9.1 shows that if text is to be added as shown in the two illustrations, the text style obliquing angle and the text rotation angle must be set if the text is to be aligned with the isoplane. Text entered when in an isometric snap style, does not automatically assume the isometric angle.

The dialogue box Drawing Aids

Ortho, Grid and Snap can be set by selecting **Drawing Aids ...** from the **Settings** pull-down menu and entering and selecting settings within the boxes of the dialogue box. In the illustration Fig. 9.3 it

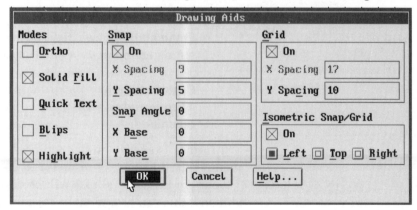

Fig. 9.3 The **Drawing Aids** dialogue box showing **Isometric Snap** on

Pictorial drawing 153

will be noted that Ortho is off (its box is not checked); Snap is on and set to a Y spacing of 5; Grid is set with a Y spacing of 10; Isometric Snap is set on (its box is checked). The X setting of both Snap and Grid are greyed out. This is because with Isometric Snap on, the X spacing is determined by the setting of the Y spacing. The X spacing with Isometric Snap on cannot be set. It is probably easier and quicker to set Ortho, Snap, Grid and Isometric Snap from the **Drawing Aids** dialogue box than to set them from the keyboard.

Examples of isometric drawings

Three simple isometric drawings are shown in Fig. 9.4 When constructing this type of drawing:

1. **SNAP** style must be set to **Isometric**;
2. If **GRID** is on, it is easier to see the isometric axes. It is often easier to construct a pictorial drawing with **GRID** set to a greater unit size than **SNAP** – e.g. grid set to 10 and snap set to 5;
3. Make full use of snap points and, at times, **osnaps**;
4. If **ORTHO** is **ON**, when lines and polylines are being drawn, any movement of the selection device (usually a mouse) causes the movement to be forced along the snap point lines depending upon the setting of the **Isoplane** directions. This makes the drawing of lines along the isometric axes easier;
5. Similarly if **COORDS** is on and showing relative coordinate

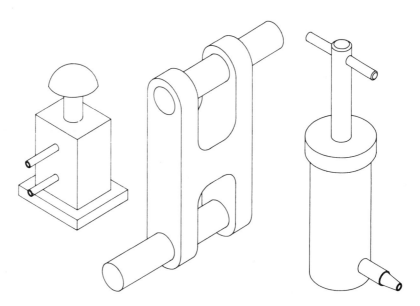

Fig. 9.4 Three examples of simple isometric drawings constructed in AutoCAD

lengths, the actual unit lengths of lines along the isometric axes can be more easily set;
6. Use the commands **MOVE** and **COPY** to speed up drawing – particularly of isometric circles (ellipses). Repeated copying of an ellipse along the snap axes may often be necessary in this form of drawing.

Figure 9.5 shows an exploded isometric drawing – a tool holding device from a shaper machine. Note that in this drawing all outer lines have been drawn with plines, to act as a form of shading for the various parts of the device and to accentuate their shapes.

Fig. 9.5 An exploded isometric drawing constructed in AutoCAD

Exercises

Exercise 1

Figure 9.6 – drawing 1. Construct an isometric drawing of the item shown by two orthographic views. Do not include dimensions in your answer.

Exercise 2

Figure 9.6 – drawing 2. Construct an isometric drawing of the item shown by two orthographic views. Do not include dimensions in your answer.

Pictorial drawing

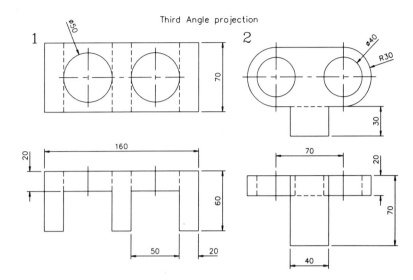

Fig. 9.6 Exercises 1 and 2

Exercise 3

Figure 9.7 – drawing 3. The two given views show a fork connector. Construct an isometric drawing of the connector. Do not include dimensions with your drawing.

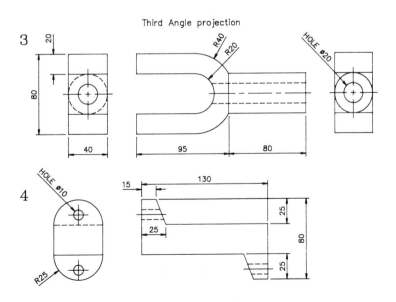

Fig. 9.7 Exercises 3 and 4

Exercise 4

Figure 9.7 – drawing 4. Construct an isometric drawing of the item shown by front and end views. Do not include dimensions.

Exercise 5

Figure 9.8. This is a much more difficult exercise and is in three parts:

(i) Construct isometric drawings of each of the parts of the assembly.
(ii) Construct an isometric drawing of the assembly with all parts in their final positions relative to each other. Try using your previously drawn parts drawings by inserting them into a final drawing, exploding the inserts and trimming unwanted lines.
(iii) Construct an exploded isometric drawing with all parts drawn as if pulled straight out of their holes. This may also be drawn by insertion of your previously drawn parts drawings.

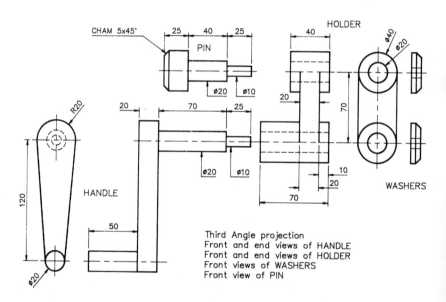

Fig. 9.8 Exercise 5

Isometric drawing is not a three-dimensional method

As mentioned in the introduction to this chapter, isometric drawing is not truly three-dimensional (3D), even though on paper it appears to be so. In a later chapter, when we are dealing with true 3D drawing, it will be seen that a command **VPOINT** allows the operator to look at 3D drawings from a variety of angles. To

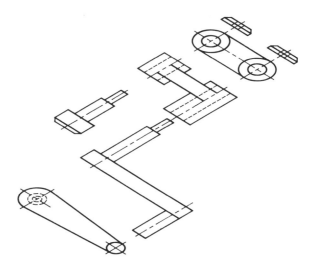

Fig. 9.9 Isometric drawing is not three-dimensional

illustrate that isometric drawing is not 3D, Fig. 9.9 is a VPOINT view of the exploded orthographic drawing Fig. 9.8. As can be seen, the drawing is lying flat on a two-dimensional (2D) plane.

Oblique drawing

Another pictorial drawing method which can be practised with the aid of AutoCAD is oblique drawing. The simplest procedure for this form of drawing is:

1. Construct a front view of the item being drawn;
2. Draw lines at an angle (normally 45 degrees), from suitable points on the front view;
3. Repeat the front view at the ends of the 45-degree lines.

This method of pictorial drawing is reasonably well suited to items which have a fairly complicated front view coupled with a simple end view. It does however have its drawbacks, the obvious one being that the resulting drawing usually looks somewhat distorted. This can be overcome to some extent by constructing what is known as a cabinet drawing. This entails the same procedures as for oblique drawing with the exception that measurements along the 45-degree lines are taken at a scale of less than one – often at half size.

If **GRID** and **SNAP** are set with both X and Y units the same, the 45-degree angles are easily obtained by drawing lines across from

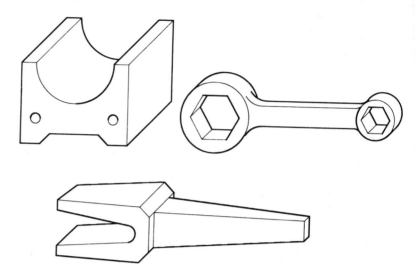

Fig. 9.10 Three examples of oblique projection

corner to corner of the grid pattern from suitable points on the front view. As the method is not all that suitable for use with AutoCAD 12, only three simple examples of oblique drawing are given in Fig. 9.10. Each of these three examples has had pline outlines in an attempt to produce a 3D effect to the drawings.

Fig. 9.11 An example of planometric drawing

Planometric drawing

Planometric drawing involves first drawing a plan of the item being illustrated. This is drawn at angles of either 45 degrees or 30/60 degrees to the horizontal. In AutoCAD these angles are obtained by using the **Rotate** option of the **Snap** command sequence. If a 45/45 degree set-up is required the rotation will be at 45; if a 30/60 rotation is required it will be at 30. When the **Snap/Grid** is rotated the square pattern of snap/grid points remains, but set at the angle stipulated. This makes it easy, with the aid of **ORTHO** and **COORDS**, to draw the required plan, from suitable points on which verticals are taken to obtain the necessary illustration.

Figure 9.11 is an example of a planometric drawing derived from a plan set at the 30/60 angles. Other examples suitable for this form of drawing would be room and building layouts, exhibition layouts, some street layouts and similar types of illustration.

CHAPTER 10

The 3D Surfaces commands

Introduction

The **Surfaces** commands, together with the command **ELEVATION**, can be employed for constructing 3D drawings (often referred to as 3D models or 3D solids). However, with the introduction of the Advanced Modelling Extension (AME) with AutoCAD Release 11 and with its latest modifications (in Release 2.1 of AME) in AutoCAD 12, the **Surfaces** commands are not as important as they were prior to the introduction of AME. Because of this, the contents of this chapter contain only very brief descriptions of the **Surfaces** commands. A more detailed explanation of the construction of 3D solids will be given in the chapter dealing with AME (pages 172 to 211).

The AutoCAD 3D coordinate system

So far in this book we have been dealing with two-dimensional drawing (2D), with the two coordinate axes x and y. In fact any point in AutoCAD can be determined with relation to three coordinate axes x, y and z. The position of points can be given in numbers which relate to the coordinate origin x,y,z = 0,0,0, usually placed at the bottom left of the AutoCAD drawing editor screen. The 3D axes lie as shown in Fig. 10.1:

+ve x is horizontally to the right of origin 0,0,0;
−ve x is horizontally to the left of the origin 0,0,0;
+ve y is vertically (on the screen) above the origin 0,0,0;
−ve y is vertically (on the screen) below the origin 0,0,0;
+ve z is perpendicular to the screen surface from the origin 0,0,0 and outwards towards the operator;
−ve z is perpendicular to the screen surface from the origin 0,0,0 and inwards away from the operator.

If working, and then plotting or printing to a scale of full size

The 3D Surfaces commands

with **LIMITS** set to 420,297 (A3 sheet sizes), then each coordinate unit can be taken as representing 1 millimetre. This also applies when working with **LIMITS** set to other metric sheet sizes. If working in other units – e.g. feet, or inches and working full size – each coordinate point can be assumed as being one unit of measurement, providing **LIMITS** is suitably set.

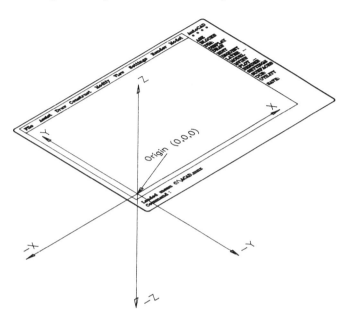

Fig. 10.1 The 3D AutoCAD axes

3D Surfaces commands from the Draw pull-down menu

The **Draw** pull-down menu is illustrated in Fig. 10.2 with the cascaded **3D Surfaces** sub-menu. The items of this sub-menu display the commands:

Edgesurf – for drawing 3D surfaces defined by four edges, which must be joined to each other's ends;
Rulesurf – for drawing 3D surfaces defined by two lines and/or arcs;
Revsurf – for drawing 3D solids of revolution around a defined pline;
Tabsurf – for drawing extrusions from pline outlines;
3dface – for drawing triangular or quadrilateral plane surfaces in 3D;
3dobjects – when this command is entered or selected from the sub-menu, the **3D Objects** dialogue box appears on screen.

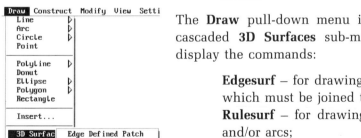

Fig. 10.2 The **Draw** pull-down menu and the cascaded **3D Surfaces** sub-menu

Fig. 10.3 The **View** pull-down menu with the cascaded **Set Views** sub-menu

The command 3DFACE

3DFACE can be entered as a command at the command line, selected from the **Draw** pull-down menu, or selected from the **DRAW** on-screen menu. As with all other commands in AutoCAD 12, no matter which method of calling the command is used, the on-screen menu area changes to show the options available with the command. When **3DFACE** is entered at the command line, the on-screen menu changes to give the options shown in Fig. 10.3 and a series of prompts appear at the command line requesting that the operator selects points on screen for each corner of the 3dface. As the command is concerned with three-dimensional constructions, coordinate numbers are frequently required when the command is in operation. The following sequence of options and responses is based on the left-hand face of the rectangular block in Fig. 10.4.

Command: 3dface   *right-click*
First point: 70,200   *right-click*
Second point: 70,100   *right-click*
Third point: 70,100,100   *right-click*
Fourth point: 70,200,100
Third point: *right-click*
Command:

Notes

1. 3dfaces can only be drawn with three or four edges;
2. **Third point:** appears after the final **Fourth point:** of the face is given. A *right-click* in response completes the face;
3. Another face can be constructed by continuing with the selection of a **Third point** at the end of the given sequence;
4. Selecting the option **.xy** from the on-screen menu for the

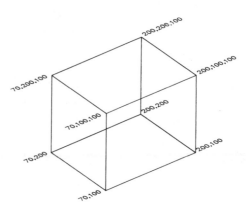

Fig. 10.4 A rectangular block drawn with the aid of the command **3DFACE**

command, or entering **.xy** from the keyboard, allows the x and y coordinate points to be *picked* by pointing, with the z component of the 3D coordinate number to be entered at the keyboard. The method would be:

Command: 3dface   *right-click*
First point: 70,200   *right-click*
Second point: 70,100   *right-click*
Third point: **.xy**   *pick* the point 70,100 **(need Z)** 100
                                                              right-click
Fourth point: **.xy**   *pick* the point 70,200 **(need Z)** 100
                                                              right-click
Third point: *right-click*
Command:

The command HIDE

When this command is called, all lines behind 3dfaces are hidden. Its use involves entering only the command followed by a *right-click*:

Command: hide   *right-click*
Regenerating drawing ...
Hiding lines: 100%
Command:

Figure 10.5 shows two seemingly identical rectangular blocks. The left-hand one was constructed with 3dfaces, the right-hand one was constructed with lines. Calling **HIDE** resulted in all lines behind the front and right-hand faces becoming hidden.

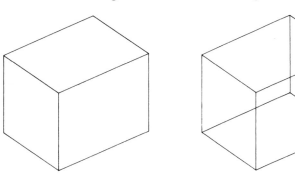

Fig. 10.5 The effect of the command **HIDE**

The command VPOINT

This command allows 3D drawings to be viewed from a variety of positions in 3D space. The command can be entered from the

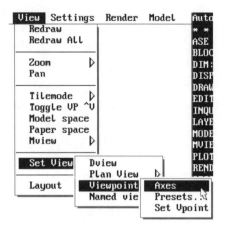

Fig. 10.6 The **3DFACE** on-screen menu

keyboard or **Viewpoint** can be selected from the **View** pull-down menu. If entered at the keyboard, the sequence of options at the command line will be:

Command: vp (vpoint)   *right-click*
Rotate/<Viewpoint><0,0,1>: −1,−1,1   *right-click*
Command:

and the 3D drawing appears on screen in its new viewing position (Fig. 10.6).

If the response is **r** (Rotate), figures of angles for two viewing directions will be requested.

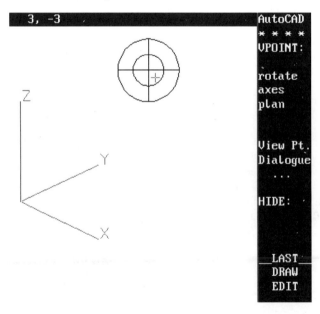

Fig. 10.7 The **VPOINT** World and tripod icons

The 3D Surfaces commands

If the response is a double *right-click* or double *Return*, the screen changes to show the icons as in Fig. 10.7. The tripod icon shows the positions of the three coordinate axes for the positioning of the 3D drawing. In the "World" icon the small cross is a cursor which can be moved with the aid of the selection device. As the cursor is moved, so the axes of the tripod icon move in response, indicating the positions which the drawing will assume. Another *right-click* brings the drawing back on screen in the viewing position indicated by the three axes of their tripod icon.

The figures entered in response to the **VPOINT** command indicate the direction from points on each of the axes looking towards the origin 0,0,0. The figures only represent comparisons between the three figures for x, y and z along the axes – they do not represent a distance along the axis. Figure 10.8 shows four **VPOINT** views of a 3D drawing. These four views show:

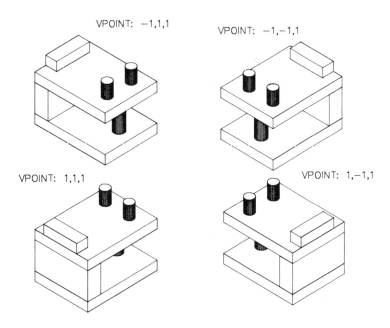

Fig. 10.8 Four **VPOINT** viewing positions

1. That labelled VPOINT: −1,1,1 is looking at the front, left and top;
2. That labelled VPOINT: −1,−1,1 is looking at the front, right and top;
3. That labelled VPOINT: 1,1,1 is looking at the back, left and top;
4. That labelled VPOINT: 1,−1,1 is looking at the back, right and top.

If the z component of the x,y,z viewpoint figures is negative, the resulting view will be as if looking from below.

VPOINT can also be employed to copy and position 3D drawings in orthographic projections. Figure 10.9 is an example. The 3D drawing was filed as three separate wblocks, each viewed from vpoints as shown, then inserted into a new drawing. Note this method is more suited to forming orthographic projections from AME solids in viewports (see later chapters).

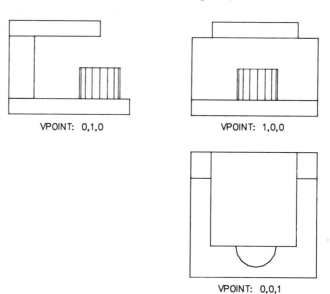

Fig. 10.9 A 3D drawing copied and inserted to form an orthographic projection

The 3D Surfaces commands

The four commands **EDGESURF**, **RULESURF**, **REVSURF** and **TABSURF** will form 3D four-sided surface patches within areas defined by entities drawn in 3D. The densities of the surface patches are controlled by two variables **Surftab1** and **Surftab2**. Except when employing **EDGESURF**, the setting for **Surftab2** is usually 2.

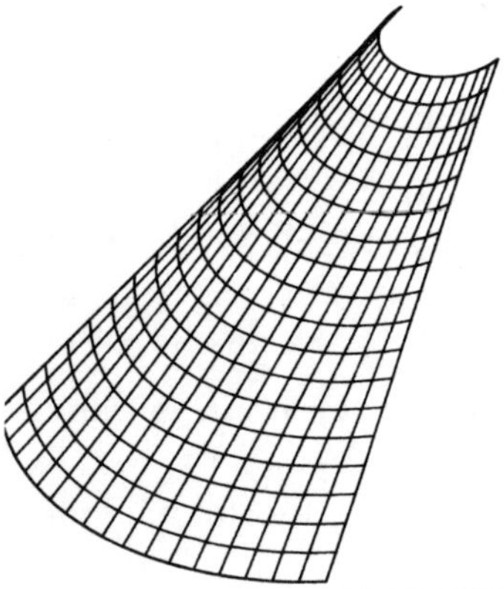

The command EDGESURF

Fig. 10.10 A **VPOINT** view of the surface defined in Fig. 10.11

An example of the use of this command is given in Figs 10.10 and 10.11. The surface generated by using the command depends upon adjoining line and/or arc entities. If the four entities do not meet at their ends, the surface will not form. Both **Surftab** variables need to be set at a value higher than 2. When entered from the keyboard the command line shows:

The 3D Surfaces commands 167

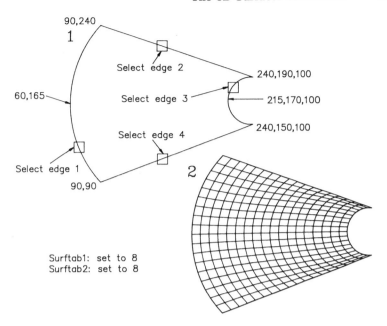

Fig. 10.11 An example of the use of **EDGESURF**

Command: edgesurf *right-click*
Select edge 1: *pick*
Select edge 2: *pick*
Select edge 3: *pick*
Select edge 4: *pick*
Command:

and the surface forms. If the ends of the four entities surrounding the surface do not meet, a warning such as **Edge 3 does not touch another edge.** appears. The four entities need reparing before the action can be completed.

The command RULESURF

Fig. 10.12 A **VPOINT** view of the surface defined in Fig. 10.13

To form a surface patch between two 3D entities, **Surftab2** should be set at 2. Figures 10.12 and 10.13 show an example of the surfaces formed with the use of the command. After drawing the two entities for the surface, enter the command:

Command: rulesurf *right-click*
Select first defining curve: *pick*
Select second defining curve: *pick*
Command:

and the surface appears between the two entities. Note that the manner in which the surface forms depends upon which half of

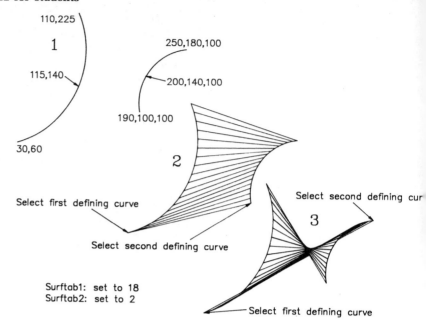

Fig. 10.13 Surfaces formed with the command **RULESURF**

Surftab1: set to 18
Surftab2: set to 2

either entity is *picked*. This is shown in drawings 2 and 3 of Fig. 10.13 and by the **VPOINT** views of the two surfaces in Fig. 10.12.

The command REVSURF

The **Surftab2** setting is again 2. Figure 10.14 shows an example of a surface formed with the aid of the command. When used, the command enables a 3D surface in the form of a solid of revolution from a pline outline around an axis. Either a full circle of revolution or a partial circle can be formed. The command line shows:

> **Command:** revsurf   *right-click*
> **Select path curve:**   *pick*
> **Select axis of revolution:**   *pick*
> **Start angle<0.00<:**   *right-click* (to accept the angle of 0)
> **Included angle (+=ccw, −=cw)<Full circle>:**   *right-click*
> (to accept full circle)
> **Command:**

and the solid of revolution forms.

The command TABSURF

Again, the **Surftab2** setting for this command should be 2. The

The 3D Surfaces commands   169

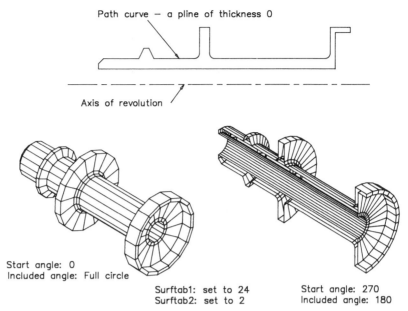

Fig. 10.14 Examples of the use of the command **REVSURF**

command will extrude a pline outline in line with an entity set anywhere near the pline outline. Fig. 10.15 shows an example of a **TABSURF** 3D drawing. The command sequence is:

Command: tabsurf   *right-click*
Select path curve: *pick*
Select direction vector: *pick*
Command:

and the extrusion appears on screen. Note that if the upper end of the direction vector is *picked*, the extrusion takes place below the pline outline.

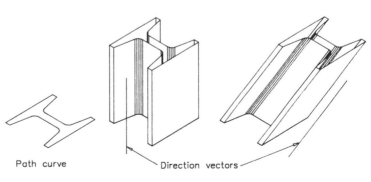

Fig. 10.15 An example of the use of **TABSURF**

The command ELEVATION

This command allows outlines drawn with 2D entities to be extruded in the direction of the z axis. When the command is called the command line shows:

Command: elev (elevation)   *right-click*
New current elevation <0>:   *right-click* (to accept 0)
New current thickness <0>: 100   *right-click* sets the extrusion thickness above the x,y plane to 100 units in the z direction
Command:

Figure 10.16 shows a number of entities drawn on the x,y plane, each drawn with the current elevation at 0, but with various elevation thicknesses. Figure 10.17 shows the 3D extensions viewed with VPOINT at −1,−1,1. In Fig. 10.17 note the following:

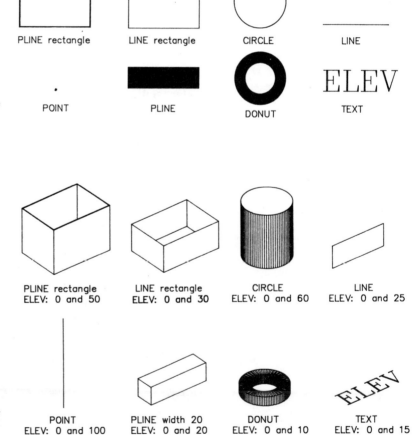

Fig. 10.16 Plan views of extruded entities lying in the x,y plane

Fig. 10.17 **VPOINT** views of the extrusions of Fig. 10.16

Pline and **Line**: the extrusions resulting from outlines drawn plines or lines will not include a 3dface at the top of the extrusion;
Circle: an extruded circle does have a 3dface on its top;
Line: a single line results in a vertical 3dface;
Point: a point extrudes as a line;
Pline: a pline of greater width than 0 will extrude as a completely enclosed 3D drawing, with a 3dface on its top;
Donut: a donut extrudes as a ring with a number of adjoining 3dfaces;
Text: will not extrude.

CHAPTER 11

The Advanced Modelling Extension (AME)

Introduction

AME is an extension program to AutoCAD 12. If the extension files have been installed with the AutoCAD files, when any AME command is called, AME automatically loads into memory. AME is a solid modelling program which allows 3D solid models to be constructed in AutoCAD 12. The solid modelling commands can be called by selection from the **Model** pull-down menu (Fig. 11.1), or by entering the command name at the keyboard. Although the *acad*.pgp file (page 12) includes abbreviations for the AME commands, in this book, when AME command names are given they will always include the prefix **SOL**.

The AME primitives

Select **Primitives** from the **Model** pull-down menu and the **AME Primitives** dialogue box appears on screen. As Fig. 11.2 shows, the six primitive solids can be seen in this dialogue box. Any one of these can be selected by a *double-left* on the icon of the required solid, or by a *left-click* on the icon followed by a *left-click* on **OK**. As a primitive is selected, a triple rectangle appears around the icon showing the selection has been made.

Each solid primitive has its own set of options showing at the command line. When any of the AME commands is in operation, once all responses to the options have been given, a sequence of statements will appear at the command line, such as:

> Phase I – Boundary evaluation begins.
> 2 of 9 of Phase I in process.
> Phase II – The tessellation computation begins.
> 2 of 6 of Phase II in process.
> Updating the Advanced Modeling Extension database.
> Command:

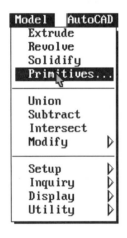

Fig. 11.1 The **Model** pull-down menu

The Advanced Modelling Extension (AME)

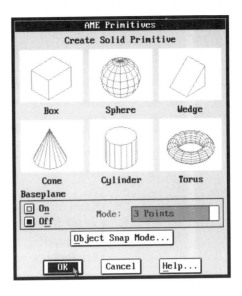

Fig. 11.2 The **AME Primitives** dialogue box

These sequences are not included in any of the AME commands options and responses given below.

When a primitive command is called the command line shows:

SOLBOX
Command: solbox   right-click
Baseplane/Center/<Corner of box><0,0,0>:   pick (or
  enter x,y,z   right-click)
Cube/Length/<Other corner>:   pick (or enter x,y,z   right-click)
Height: enter a number (the z coordinate)   right-click
Command:
SOLSPHERE
Command: solsphere   right-click
Baseplane/<Center of sphere><0,0,0>:   pick (or enter
  x,y,z   right-click)
Diameter/<Radius>of sphere: pick (or enter a number
  right-click)
Command:
SOLWEDGE
Command: solwedge   right-click
Baseplane/<Corner of wedge><0,0,0>:   pick (or enter
  x,y,z   right-click)
Length/<Other corner>:   pick (or enter x,y,z   right-click)
Height: enter a number   right-click
Command:

SOLCONE
Command: solcone   *right-click*
Baseplane/Elliptical/<Center point><0,0,0>:   *pick (or enter x,y,z   right-click)*
Diameter/<Radius>:   *pick (or enter a number   right-click)*
Apex/<Height>: enter a number   *right-click)*
Command:
SOLCYL
Command: solcyl   *right-click*
Baseplane/Elliptical/<Center point><0,0,0>:   *pick (or enter x,y,z   right-click)*
Diameter/<Radius>:   *pick (or enter a number   right-click)*
Center of other end/<height>: enter a number   *right-click*
Command:
SOLTORUS
Command: soltorus   *right-click*
Baseplane/<Center of sphere><0,0,0>:   *pick (or enter x,y,z   right-click)*
Diameter/<Radius> of torus: *pick (or enter a number   right-click)*
Diameter/<Radius>of tube: *pick (or enter a number   right-click)*

Command:

Because of space limitations in a book of this type, only the default options (those given in brackets < >) will be considered in the examples outlined below. The reader is, however, advised to experiment with other AME primitive options in order to appreciate how effective the AME modeller can be. Figure 11.3 is a VPOINT view of some AME primitives, together with the alternatives available with **BOX** (Cube), **CONE** (Elliptical) and **CYLINDER** (Elliptical).

The AME Boolean operators

AME solids, including those formed with **SOLEXT**, **SOLREV** (page 180) can be joined, subtracted from each other or intersected with each other using the command systems **SOLUNION** (union), **SOLSUB** (sub) and **SOLINT**. These commands cause the Boolean operations of union, difference and intersection to take effect on any AME solid models.

The Advanced Modelling Extension (AME) 175

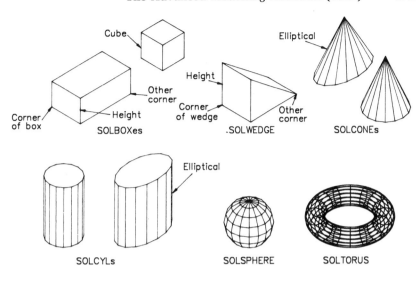

Fig. 11.3 Examples of AME primitives

SOLWDENS set at 4   MESH except for TORUS   HIDE except for TORUS

When called, these three commands produce the following sequence of options at the command line. Following the final *right-click*, a series of statements similar to those for the primitive commands appear at the command line, indicating to the operator that he/she must wait until the computation for the Boolean action is completed:

Command: solunion (or union)   *right-click*
Select objects: pick **1 selected, 1 found**
Select objects: pick **1 selected, 1 found**
Select objects: *right-click*
Command:

The **SOLSUB** (sub) and **SOLINT** show a similar set of options.

Notes

1. When any AME primitives have been acted upon by the Boolean operators, the resulting solids can be separated into their primitives by using the command **SOLSEP**. The action of this command is to break up a combined, separated or intersected solid into its constituent primitives;
2. Before the action of **HIDE** can take effect on an AME solid, it must first have its wireframe structure changed into a series of surface meshes. This is effected by the action of the command **SOLMESH** (mesh).

Examples of Boolean operations in AME

Figure 11.4 compares the actions of the three Boolean operators.

Drawing 1

Although the drawing shows the torus and cylinder separate from each other, in fact if they are drawn to the coordinates given with the drawing, the torus will be halfway up the cylinder.

Drawing 2

The two solids after the action of **SOLUNION**, **SOLMESH** and **HIDE**.

Drawing 3

Call **SOLSEP** and separate the union of the two solids, then call **SOLINT**, followed by **SOLMESH** and **HIDE**.

Drawing 4

Call **SOLSEP** again to separate the subtraction, then call **SOLSUB**, **SOLMESH** and **HIDE**.

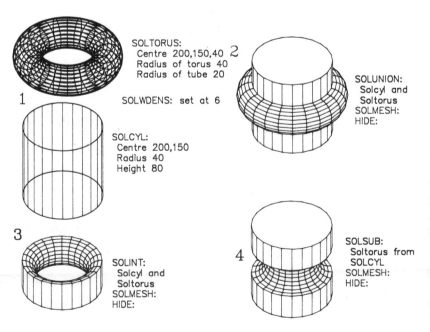

Fig. 11.4 Examples of solids produced with the three Boolean operators

Examples of AME solid models

Figure 11.5 illustrates a sequence for constructing a simple AME solid model from two primitives with the aid of the two Boolean operators union and subtract. Note the statement with Fig. 11.5 that the variable **SOLWDENS** is set to 6 for the model shown. This variable can be set from between 2 and 16 and determines the number of surface meshes which will be contained in the circular surfaces of primitive cylinders or cones.

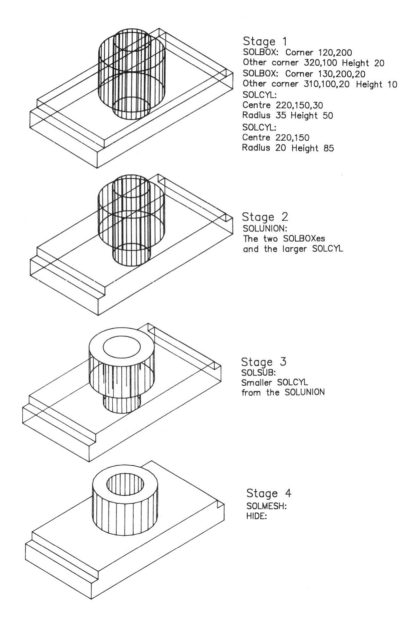

Fig. 11.5 Stages in constructing a simple AME solid model

The commands SOLFILL and SOLCHAM

Enter solfill at the keyboard and the command line shows:

> **Command:** solfill   *right-click*
> **Pick edges of solids to be filleted (press ENTER when done):**   *pick pick right-click*
> **2 edges selected:**   *right-click*
> **Diameter/<Radius>of fillet<10.00>:** 5   *right-click*

series of statements finishing with

> **Updating the Advanced Modelling Extension database.**
> **Command:**

The selected edges are then filleted (or radiused) to the unit size given as a response to the option

> **Diameter/<Radius>of fillet.**

Note: It is advisable to place the solid, the edges of which are to be filleted, in a **VPOINT** view − say either vpoint 1,1,1 or vpoint −1,−1,1. Unless this is done it is practically impossible to select the appropriate edge(s) which are to be acted upon. **ZOOM** windows may also be necessary to locate some details.

Figure 11.6 is an example of a solid with two edges acted upon with the command. The reader is advised to work the solid model shown in Fig. 11.6 − all necessary commands and coordinates are included in the illustration.

Enter solcham at the keyboard and the command line shows:

> **Command:** solcham   *right-click*
> **Pick base surface:**
> **Next/<OK>:**

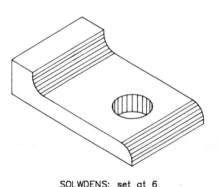

Fig. 11.6 A solid model with filleted edges

SOLWDENS: set at 6

```
SOLBOX:
    Corners 60,200 and 90,100
    Height 40
SOLBOX:
    Corners 90,200 and 230,100
    Height 20
SOLCYL:
    Centre 180,150
    Radius 15 Height 25
SOLUNION:
    The two SOLBOXes
SOLSUB:
    SOLCYL from the SOLUNION
SOLFILL:
    Radius 15
    Pick the two edges to be filleted
SOLMESH:
HIDE:
```

The Advanced Modelling Extension (AME) 179

Pick edges of face to be chamfered (press ENTER) when done): *pick right-click*
Enter distance along base surface <0.00>: 5 *right-click*
Enter distance along adjacent surface <10.00>: *right-click*

series of statements finishing with
Updating the Advanced Modeling Extension database.
Command:

Notes

1. After picking a base surface it highlights. If it is not the correct base surface at the first *pick*, instead of responding to **Next/<OK>:** with a *right-click*, enter n (Next) and an adjacent surface to that selected will highlight. If necessary, continue with requesting the Next surface until the required surface highlights – the required surface is that which contains one corner edge of the chamfer;
2. As with constructing a fillet, when chamfering in AME, it is advisable to place the model in a **VPOINT** view in order to facilitate selection of the required base surfaces;
3. Although the distance along the adjacent edge to that first selected will be accepted as a default distance for the adjacent edge, a chamfer can be constructed with different distances along the two surfaces.

Figure 11.7 is an example of a simple solid model with chamfered edges. Again, the reader is advised to work this example to understand how the **SOLCHAM** command system functions.

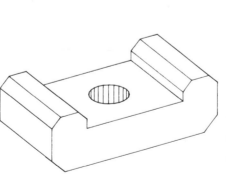

```
SOLBOX:
    Corners 100,200 and 140,100
    Height 60
SOLBOX:
    Corners 250,200 and 290,100
    Height 60
SOLBOX:
    Corners 140,200 and 250,100
    Height 40
SOLCYL:
    Centre 195,150 Radius 20
    Height 45
SOLUNION:
    The three SOLBOXes
SOLSUB:
    SOLCYL from SOLUNION
SOLCHAM:
    Distances 10 and 10
    and Distances 15 and 15
SOLMESH:
HIDE:
```

SOLWDENS: set at 6

Fig. 11.7 An AME solid model with chamfered edges

The commands SOLEXT and SOLREV

The command **SOLEXT** will form solid models by extrusions from outlines constructed from polylines, arcs or circles. The command **SOLREV** will form solids of revolution around an axis from outlines constructed from polylines, arcs or circles.

When solext is entered at the keyboard, the options appearing at the command line are:

> **Command:** solext   *right-click*
> **Select regions, polylines and circles for extrusion ...**
> **Select objects:**   *pick* **1 found**
> **Select objects:**   *right-click*
> **Height of extrusion:** 50   *right-click*
> **Extrusion angle <0>:**   *right-click*
> **Command:**

Note: as with solfill and solcham, it is advisable to place the pline outline being extruded into a **VPOINT** view in order to make selections easier.

Figure 11.8 shows a solid model formed mainly from an extrusion. The reader is advised to follow the instructions given

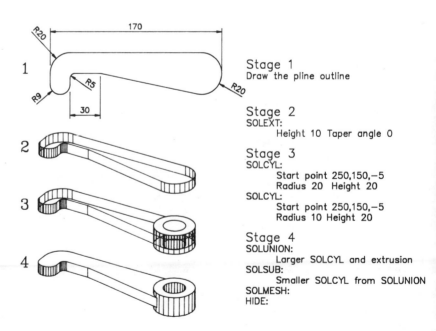

Fig. 11.8 An example of an extrusion constructed with the aid of **SOLEXT**

with this illustration to learn the methods of constructing this type of solid model.

Call the command **SOLREV** and the command line shows:

Command: solrev   *right-click*
Select region, polyline or circle for revolution ...
Select objects: *pick* **1 found**
Select objects: *right-click*
Axis of revolution — Entity/X/Y/<Start point of axis>:
                              pick   **End point of axis:**   *pick*
Angle of revolution <full circle>:   *right-click*
Command:

and the solid of revolution forms.

Notes

1. The axis of revolution need not be drawn;
2. The axis of revolution must be a straight line or pline;
3. The axis of revolution may not necessarily be parallel to the main angle of the outline from which it is to be formed;
4. The angle of revolution can be of any size (less than 360 degrees) to produce a half, a quarter, etc., solid of revolution.

Two examples of solids of revolution, constructed with the aid of **SOLREV**, are given in Figs 11.9 and 11.10. Note that in Fig. 11.10, a hole forms in the solid if the pline outline from which the solid is derived does not meet the axis of revolution.

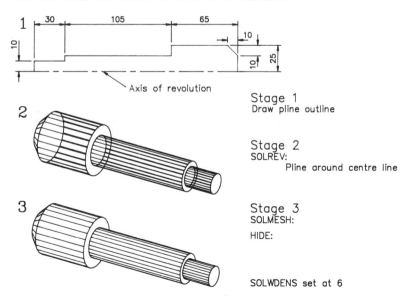

Fig. 11.9 An example of a solid of revolution constructed with the aid of the command **SOLREV**

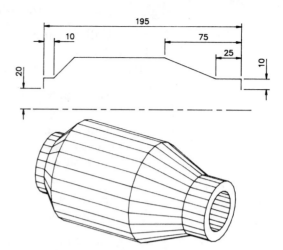

Fig. 11.10 Another example of a solid of revolution

Examples of AME solid models

Figure 11.11 shows four examples of relatively simple solid models constructed with the aid of the commands so far demonstrated in this chapter. The reader may wish to attempt constructing these four examples (or similar simple objects) working to any dimensions thought to be suitable.

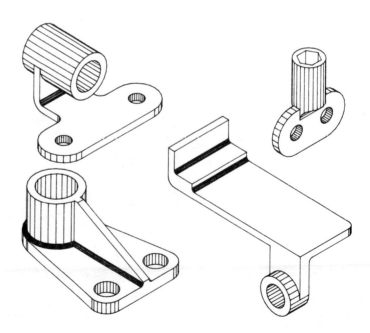

Fig. 11.11 Four examples of AME solid models

The Advanced Modelling Extension (AME)

Exercises

Exercise 1

Figure 11.12, drawing 1. Working to the details given with drawing 1 construct the solid model.

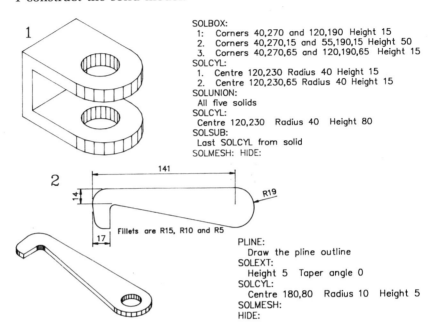

Fig. 11.12 Exercises 1 and 2

SOLBOX:
1: Corners 40,270 and 120,190 Height 15
2. Corners 40,270,15 and 55,190,15 Height 50
3. Corners 40,270,65 and 120,190,65 Height 15
SOLCYL:
1. Centre 120,230 Radius 40 Height 15
2. Centre 120,230,65 Radius 40 Height 15
SOLUNION:
All five solids
SOLCYL:
Centre 120,230 Radius 40 Height 80
SOLSUB:
Last SOLCYL from solid
SOLMESH: HIDE:

PLINE:
  Draw the pline outline
SOLEXT:
  Height 5 Taper angle 0
SOLCYL:
  Centre 180,80 Radius 10 Height 5
SOLMESH:
HIDE:

Exercise 2

Figure 11.12, drawing 2. Working to the details given with drawing 2 construct the solid model.

Exercise 3

Figure 11.13, drawing 3. Working to the details given with drawing 3 construct the solid model.

Exercise 4

Figure 11.13, drawing 4. Working to the details given with drawing 4 construct the solid model.

Exercise 5

Figure 11.14, drawing 5. Working to the details given with drawing 5 construct the solid model.

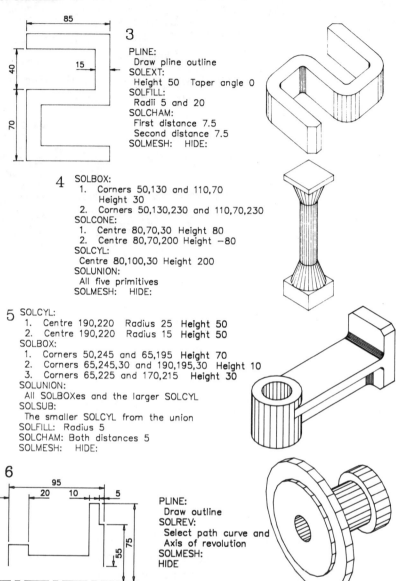

3
PLINE:
   Draw pline outline
SOLEXT:
   Height 50   Taper angle 0
SOLFILL:
   Radii 5 and 20
SOLCHAM:
   First distance 7.5
   Second distance 7.5
SOLMESH: HIDE:

4 SOLBOX:
   1. Corners 50,130 and 110,70
      Height 30
   2. Corners 50,130,230 and 110,70,230
SOLCONE:
   1. Centre 80,70,30 Height 80
   2. Centre 80,70,200 Height −80
SOLCYL:
   Centre 80,100,30 Height 200
SOLUNION:
   All five primitives
SOLMESH: HIDE:

Fig. 11.13 Exercises 3 and 4

5 SOLCYL:
   1. Centre 190,220   Radius 25   Height 50
   2. Centre 190,220   Radius 15   Height 50
SOLBOX:
   1. Corners 50,245 and 65,195   Height 70
   2. Corners 65,245,30 and 190,195,30   Height 10
   3. Corners 65,225 and 170,215   Height 30
SOLUNION:
   All SOLBOXes and the larger SOLCYL
SOLSUB:
   The smaller SOLCYL from the union
SOLFILL: Radius 5
SOLCHAM: Both distances 5
SOLMESH: HIDE:

6
PLINE:
   Draw outline
SOLREV:
   Select path curve and
   Axis of revolution
SOLMESH:
HIDE

Fig. 11.14 Exercises 5 and 6

Exercise 6

Figure 11.14, drawing 6. Working to the details given with drawing 6 construct the solid model.

Exercise 7

Figure 11.15. Working to the details given in Fig. 11.15 construct the solid model.

The Advanced Modelling Extension (AME) 185

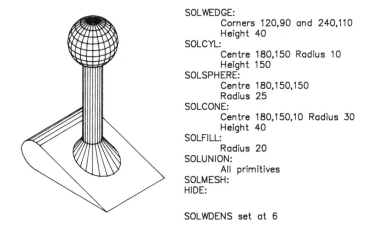

```
SOLWEDGE:
        Corners 120,90 and 240,110
        Height 40
SOLCYL:
        Centre 180,150 Radius 10
        Height 150
SOLSPHERE:
        Centre 180,150,150
        Radius 25
SOLCONE:
        Centre 180,150,10 Radius 30
        Height 40
SOLFILL:
        Radius 20
SOLUNION:
        All primitives
SOLMESH:
HIDE:

SOLWDENS set at 6
```

Fig. 11.15 Exercise 7

Further examples of AME solid models

Figure 11.16 is a solid model of a building constructed using the methods described in this chapter. The hatching of the roof in these examples requires the aid of the **UCS** (User Coordinate System) — described in the next chapter.

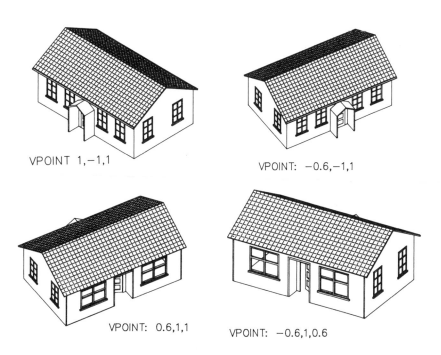

Fig. 11.16 An example of a more advanced AME solid model

Figure 11.17 shows that exploded solid models can be constructed from AME primitives. A more complex exploded solid model drawing is given in Fig. 11.18.

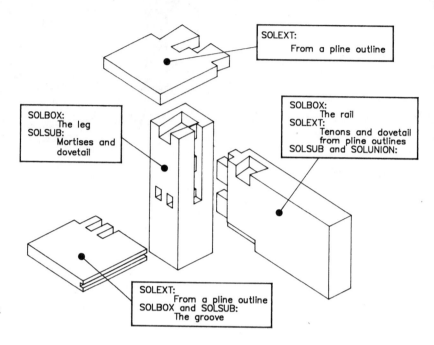

Fig. 11.17 An example of an AME exploded solid model

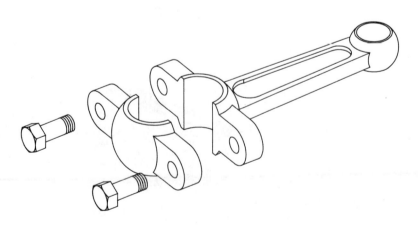

Fig. 11.18 A more advanced AME exploded solid model

The Advanced Modelling Extension (AME)   187

A more difficult exercise

Figure 11.19 is an AME solid model drawing of a garage. The hatchings of the walls and of the roof can best be added to the drawing with the aid of **UCS**, although with some ingenuity they could be added by following procedures already outlined in earlier chapters. Details of the AME commands and procedures for constructing this model are:

Snap set to 2.5
Wall SOLBOX 1: Corners 90,230 and 330,227.5
  Height 100
Wall SOLBOX 2: Corners 90,230 and 92.5,100
  Height 100
Walls COPY: Solbox 1 from 90,230 to 90,102.5
  Solbox 2 from 90,230 to 327.5,230
Roof SOLBOX 3: Corners 85,235,100 and 335,95,100
  Height 5
Roof slats SOLBOX 4: Corners 85,235,105 and 335,230,105
  Height 5
Roof slats COPY: Solbox 4 from 85,235 to 85,100
Doors SOLBOX 5: Corners 90,220 and 92.5,165
  Height 80
Doors COPY: Solbox 5 from 90,220 to 90,165
Door gap in front wall SOLBOX 6: Corners 90,220 and
                                                92.5,110
  Height 80
Window hole SOLBOX 7: Corners 130,102.5,40 and
                                                280,100,40
  Height 40
Window SOLBOX 8: Corners 130,102.5,40 and 280,97.5,40
  Height 40
Glass SOLBOX 9: Corners 135,102.5,45 and 275,97.5,45
  Height 30
SOLSUB: Solbox 7 from its wall
  Solbox 9 from Solbox 8
SOLUNION: Window and its wall
  All four walls
  Roof and its slats
  Roof, slats and walls
ROTATE: Doors by 330 and 30
SOLUNION: Doors and remainder
SOLFILL: Radius 3 between roof and its slats

Hatching is not in this exercise. Leave until **UCS** is explained later in the next chapter.

You will probably have to frequently use **VPOINT** and **ZOOM** to locate smaller selection points.

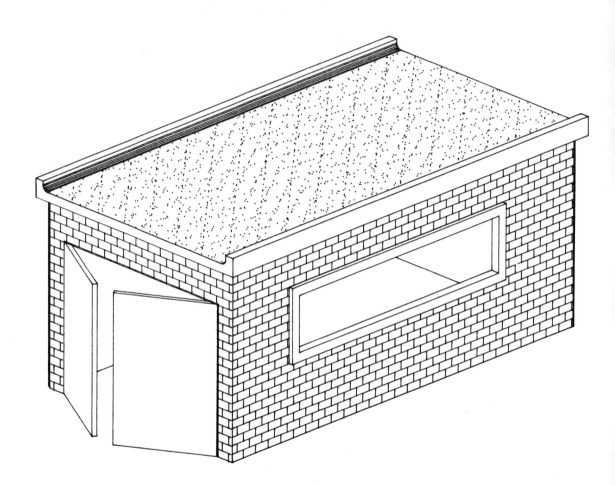

Fig. 11.19 A more difficult exercise

CHAPTER 12

UCS (User Coordinate System) and AME

Introduction

The UCS allows the operator to re-position the x,y plane on which a 3D model is being constructed to any angle or slope in 3D space. When employing the UCS, usual practice is to assume the x,y plane first set in the AutoCAD 12 drawing editor is the **WORLD** coordinate system (or WCS). Before making use of the UCS, set the two variables **UCSICON** and **UCSFOLLOW**.

    Command: ucsicon   *right-click*
    ON/OFF/All/Noorigin/ORigin/<OFF>: on   *right-click*
    Command:

and the ucsicon appears at the bottom left-hand corner of the drawing editor. As the UCS is changed, the icon takes a variety of forms – Fig. 12.1.

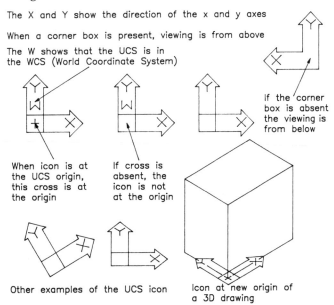

Fig. 12.1 The variety of **UCS** icons

190 AutoCAD Release 12 for students

Command: ucsfollow   *right-click*
New value for UCSFOLLOW <0>: 1   *right-click*
Command:

Unless set to 1, when a new ucs is called the change will not take effect. As will be seen later, there are occasions when it is sensible to set **UCSFOLLOW** to zero to prevent the UCS changing.

The command UCS

New positions for the UCS can be set in one of two ways:

1. Select **UCS** from the **Settings** pull-down menu (Fig. 12.2), followed by selecting **Presets ...** to bring the **UCS Orientation** dialogue box on screen (Fig. 12.3);
2. Enter the command **UCS**. The on-screen menu changes (Fig. 12.4) and the command line options and the results of making the response **3point** are:

Command: ucs (User Coordinate System)   *right-click*
Origin/Zaxis/3point/Entity/View/X/Y/Z/Prev/Restore/Save/Del/?/<World>: 3 (3point)   *right-click*
Origin point: *pick* a point in the World x,y plane
Point on positive portion of the X-axis <1,0,0>:   *pick* a second point in the World x,y plane

Fig. 12.2 The **Settings** pull-down menu

Fig. 12.4 The **UCS** on-screen menu

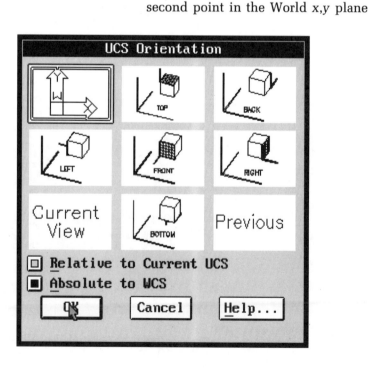

Fig. 12.3 The **UCS Orientation** dialogue box

UCS (User Coordinate System) and AME 191

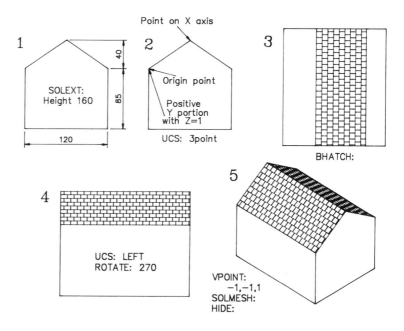

Fig. 12.5 An example of the value of **UCS**

Point on positive-Y portion of the UCS X-Y plane<0,0,1>:
.xy (entered at keyboard) **.xy of (need Z)** 1   *right-click*
Command:

As an example of constructions on a new User Coordinate System, it will be remembered that in the previous chapter Fig. 11.16 (page 185) showed hatching on the roof of a model of a building. Figure 12.5 shows how this hatching was drawn with the aid of the UCS. The five drawings of Fig. 12.5 show:

Drawing 1

The dimensions of the pline outline from which a 3D model representing the building is extruded with **SOLEXT**.

Drawing 2

The points selected for a 3point UCS in order to ensure that one side of the roof is lying in an x,y plane in a new UCS.

Drawing 3

The position of the 3D model in the 3point UCS with one side of the roof hatched with **BHATCH**. It is possible to use bhatch here because the roof is lying flat in its new x,y plane. A second 3point UCS must be called to hatch the other side of the roof.

Drawing 4

The **UCS** changed by selecting **LEFT** from the **UCS Orientation** dialogue box. The model was then rotated through 270 degrees with the aid of the command **ROTATE**.

Drawing 5

A **VPOINT** view of the model after **SOLMESH** and **HIDE**.

Figure 12.6 is a **VPOINT** view of a 3D model drawn with the aid of **AME**. Figure 12.7 shows the positions of the model in different **UCS** positions resulting from selections made from the **UCS Orientation** dialogue box.

Notes on the UCS

1. The position in which a 3D model is constructed must be considered as to how it lies in relation to the **WORLD** UCS;
2. Figure 12.6 shows a model constructed partly in the **WORLD** UCS and partly in the **FRONT** UCS. Figure 12.7 shows the model in a variety of positions as selected from the **UCS Orientation** dialogue box;
3. The UCS enables parts of 3D models to be constructed on any slope or at any angle. One problem which arises when changing the UCS is to determine where the coordinate point 0,0,0 lies. A position for the origin to be always at one corner of the model being constructed can be determined by:

Command: ucs (User Coordinate System)   *right-click*

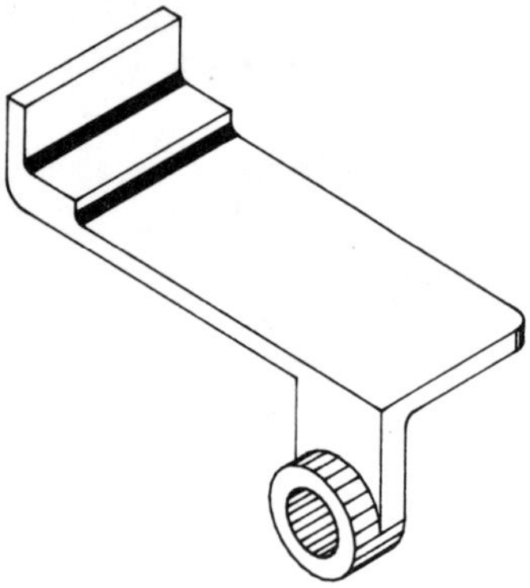

Fig. 12.6 A 3D model constructed with the aid of AME

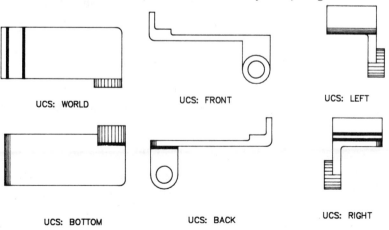

Fig. 12.7 Examples of **UCS** positions from the **UCS Orientation** dialogue box

UCS (User Coordinate System) and AME

**Origin/Zaxis/3point/Entity/View/X/Y/Z/Prev/Restore/Save/
Del/?/<World>:** o (Origin)   *right-click*
Origin point <0,0,0>:   *pick* or enter coordinates of one corner of the model
Command:

The ucsicon will settle at the corner selected by the x,y,z coordinate numbers. Figure 12.8 shows four views of a model when the ucsicon has been set with its origin at one corner of the model;

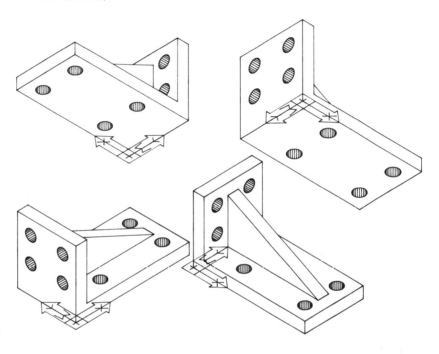

Fig. 12.8 The **UCS** icon at the Origin of four views of a 3D model

4. The **Save** and **Restore** options of the **UCS** command can be seen by selecting **UCS Control ...** from the **Settings** pull-down menu. The **UCS Control** dialogue appears with names of the saved UCS systems – Fig. 12.9.

Some more AME commands

A few more commands from the Advanced Modelling Extension (AME) will be considered in outline below. Once again it must be emphasized that the limitations of a book of this nature preclude the possibility of a full explanation of the commands within AME.

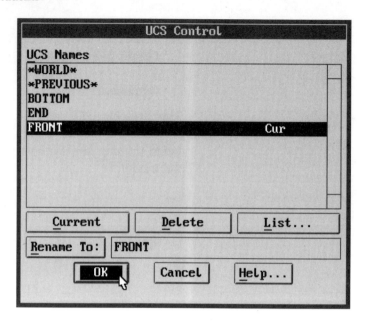

Fig. 12.9 The **UCS Control** dialogue box

The AME command SOLCUT

Any AME model can be cut into parts with the aid of **SOLCUT**. Figure 12.10 shows the results of the action of the command on a sphere. When the command is called the command line shows:

Command: solcut   *right-click*
Select objects: *pick*

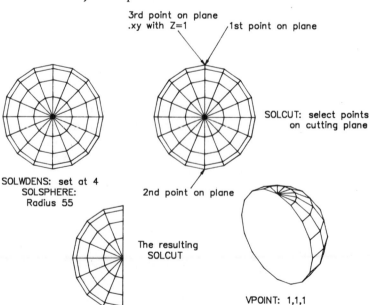

Fig. 12.10 An example of the use of the AME command **SOLCUT**

Select objects: *right-click* to confirm selection
Cutting plane by Entity/Last/Zaxis/View/XY/YZ/ZX/
<3point>: r to confirm 3point
1st point on plane: *pick*   **2nd point on plane:** *pick*
3rd point on plane: .xy   .xy of *pick*   **(need Z)** 1
right-click
Both sides/<Pick point on desired side of plane>: *pick*
series of statements showing the AME operations
Command:

Note: If the response to **Both sides/<Pick point on desired side of plane>:** is b (Both) then the sphere will be cut into two portions, but both would remain on screen.

The AME command SOLPROF

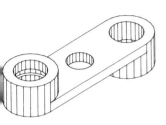

Fig. 12.11 A simple AME model

Figure 12.11 illustrates a simple AME solid model and Fig. 12.12 the same model after the action of the command **SOLPROF** has been completed. The sequences of options and prompts in the command are:

Command: solprof   *right-click*
Select objects: *pick*
Select objects: *right-click* to confirm selection
1 solid selected
Display hidden profile lines on separate layer?<Y>:
right-click to confirm Y (Yes)
Project profile lines onto a plane?<Y>: *right-click*
to confirm Y (Yes)
Delete tangential edges?<Y>: *right-click* to confirm Y
(Yes)

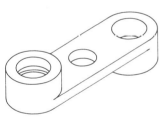

Fig. 12.12 Shows Fig. 12.11 after the action of **SOLPROF**

a series of statements giving the AME operations taking place
Command:

Notes

1. **SOLPROF** cannot operate if **Tilemode** is set to 1 (ON) – this being the normal setting of this variable. A message **This command is not available in tilemode.** will appear to prevent further operation. More about **Tilemode** later (page 000);
2. **SOLPROF** will only operate in **MSPACE** (Model Space). More about Model Space later (page 000);
3. To obtain a profile-only view of the model, select **Layer**

Control ... from the **Settings** pull-down menu and turn layer **0** and all layers with a name such as **PH-69D2** off. When these layers are turned off the profile-only outlines appear;
4. The resulting view is specific to that view only. If for example a new **VPOINT** view of the profile is called, the resulting drawing will be a distorted view.

The AME command SOLSECT

AME solid models can be section hatched with the aid of this command. The hatching pattern, spacing and angle settings are determined either by setting the solvariables as follows:

> **Command:** solhpat   *right-click*
> **Hatch pattern** <U>:   *right-click* to accept u
> **Command:**
> **Command:** solhsize   *right-click*
> **Hatch size** <1>: 4   *right-click* sets spacing to 4 units
> **Command:**
> **Command:** solhangle   *right-click*
> **Hatch angle** <45>:   *right-click* to accept angle of 45 degrees
> **Command:**

or by selecting the **System variables** from the **Model** pull-down menu. The settings can then be made from the **AME R2.1 – System Variables** dialogue box as shown in Fig. 12.13.

Figure 12.14 illustrates four stages in obtaining a section derived from the AME solid model in Fig. 12.11. The stages were:

Drawing 1

The model is constructed in a UCS **WORLD** view.

Drawing 2

The model is cut with the aid of **SOLCUT** with the front half removed.

Drawing 3

The section after the action of **SOLSECT**. The command line sequence is:

UCS (User Coordinate System) and AME   197

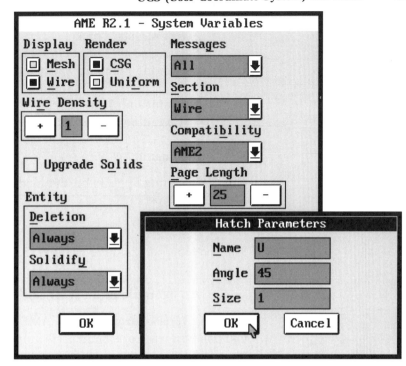

Fig. 12.13 The **AME System Variables** dialogue box

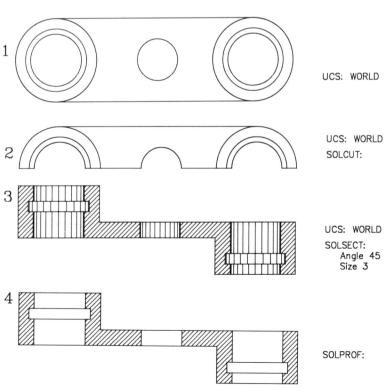

Fig. 12.14 Stages in sectioning an AME model with **SOLSECT**

Command: solsect   *right-click*
Select objects: *pick*
Select objects: *right-click* to confirm selection
**Sectioning plane by Entity/Last/Zaxis/View/XY/YZ/ZX/
              <3point>:**   *right-click* to accept 3point
1st point on plane: *pick*   **2nd point on plane:**   *pick*
3rd point on plane: .xy   .xy of   *pick*   **(need Z)** 1
                                                                          right-click

Command:

and the hatching appears on the model.

Drawing 4

The sectioned model after **SOLPROF**.

An example of an AME solid model

Figure 12.15 is an example of an AME solid model in the form of an exploded drawing of a crankshaft from a small air compressor. The model has been constructed from Solspheres; Solcylinders;

Fig. 12.15 An example of an AME solid in the form of an exploded drawing

Fig. 12.16 The model of Fig. 12.15 after the action of **SOLPROF**

Solextrusions; Solrevolutions; Solunions; Solsubs; Solcuts. The model has been Solmeshed and the resulting drawing plotted with hidden lines removed.

Figure 12.16 is the same model after the action of Solprof, with Layer 0 and the hidden lines layer set to off.

Exercises

Notes

1. In order to construct the following exercises, it will be necessary to work on a variety of User Coordinate Systems. Care needs to be taken with the positioning of primitives as the UCS is changed. Their positions in relation to each other can always be checked by calling up the **UCS Orientation** dialogue box and viewing the model from different directions. This will allow primitives which have not been placed in their correct positions to be moved as necessary. However, it is advisable to always check from two viewing directions – e.g. UCS World and UCS Front, or UCS Front and UCS Right, selecting the User Coordinate Systems from the **UCS Orientation** dialogue box;
2. After each change of the UCS, the model under construction will usually be seen occupying the full screen. Because of this, it is necessary to **ZOOM** back to scale 1 after each new UCS is called. Occasionally an All **ZOOM** may also be needed before a scale 1 zoom;
3. Remember that the UCS icon can be set at one corner of the model being constructed, using that corner as the origin;
4. There are many different ways in which any solid model can be constructed. The reader is advised to develop his/her own methods as far as is possible. The routines suggested in some of the exercises below should not be taken as the only methods of construction available with AME and UCS.

Exercise 1

Figure 12.17 is a Third Angle orthographic projection of an engineering component. Construct an AME solid model drawing of the component, working to full scale. This construction will involve the following AME and UCS commands:

SOLBOX; SOLCYL; SOLFILL; SOLUNION; SOLSUB; UCS 3point; UCS FRONT; UCS RIGHT (or LEFT); SOLMESH; HIDE.

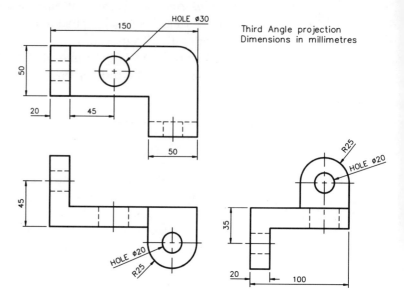

Fig. 12.17 Exercise 1

Exercise 2

Figure 12.18 suggests the stages in constructing the solid model of a bottle. Note that the outline from which the solid of revolution is formed is a continuous pline. This can be constructed by drawing a single pline, calling **OFFSET** set to 1, offsetting the original pline, followed by adding short plines at top and bottom, then adding these to the original pline with the **PEDIT** command. The AME variable controlling the density of the surfaces of the solid is **SOLWDENS**, which can be set to between 4 and 6 for this model.

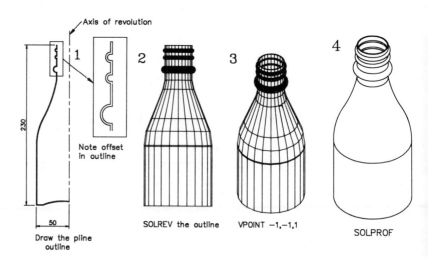

Fig. 12.18 Exercise 2

Exercise 3

Construct a bearing support following the stages given in Fig. 12.19.

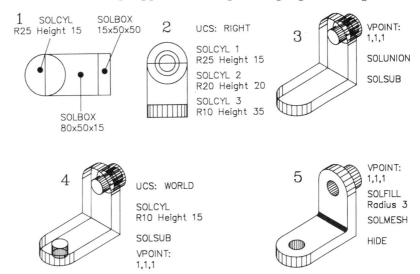

Fig. 12.19 Exercise 3

Exercise 4

Using **3D Mirror** from the **Construct** pull-down menu, mirror the bearing support constructed for Exercise 2.

Exercise 5

The 3-step pulley shown in Fig. 12.20 is a solid of revolution formed by using the command **SOLREV** on a previously drawn pline outline. When constructing the pline, make sure its axis of revolution is central to the holes in the bearing supports. The bearing supports and the 3-step pulley can be moved to correct positions after viewing in both **UCS WORLD** and **UCS FRONT**. The **VPOINT** viewing position of Fig. 12.20 is −1,−1,0.6.

Exercise 6

Having constructed the exploded model Fig. 12.20, practise the use of **SOLPROF**. The result should be as shown in Fig. 12.21.

Exercise 7

Finally construct the sectional view given in Fig. 12.22. The sequence for the construction is:

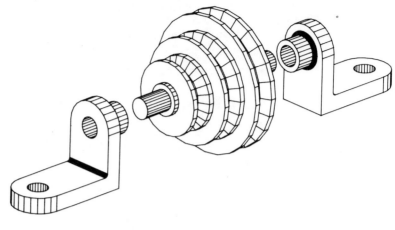

Fig. 12.20 Exercises 4 and 5

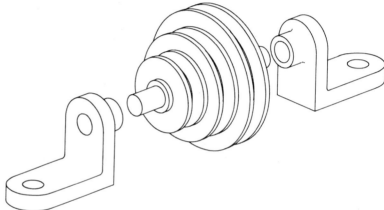

Fig. 12.21 Exercise 6

1. Ensure the model is in the **WORLD UCS**;
2. Move the three parts into their working positions in relation to each other;
3. With the aid of **SOLCUT** remove the front halves of the bearing supports. Note that the usual practice when sectioning features such as the 3-step pulley is to show them in outline – i.e. not hatched. This means the 3-step pulley should not be cut in half with **SOLCUT**;
4. Make a new layer – e.g. hatch – to take the hatching lines;
5. With the hatch pattern as u, the hatch angle as 45 and the hatch size as 5, hatch the front faces of the two bearing supports using **SOLSECT**;
6. Place the assembly in a **FRONT UCS** and move the hatch areas away from the solid models. If this is not done, the hatched areas will probably disappear when the action of **SOLPROF** takes place;

7. Call **SOLPROF** and *pick* each part of the model in turn. Turn OFF the layers **0** and those starting with **PH-**;
8. Move the hatch areas back into their correct positions.

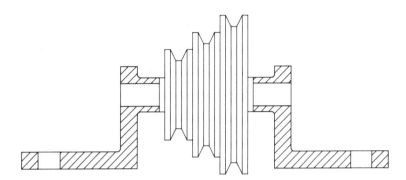

Fig. 12.22 Exercise 7

Some notes on the UCS and AME

1. After each change in the **UCS** zoom back to 1. Occasionally first zoom All before zooming to 1;
2. It is advisable for the **UCSICON** to be set at ON when constructing in AME;
3. **UCSFOLLOW** must be set to 1 for changes of the UCS to take place;
4. It is often advisable to set the UCS icon at its origin point on a corner of the model being constructed;
5. AME model drawing files tend to be large. Disk files in excess of 1 megabyte are not at all uncommon when saving AME files to disk;
6. Unless a very fast computer is in use, many of the operations in AME take a fair amount of time to complete. Even with a fast computer, the operator will have to wait for some of the operations to take place.

CHAPTER 13

Tilemode, MSpace, PSpace and Dview

Introduction

So far in this book we have been concerned with constructing drawings in **Model Space** (MSpace). 3D models can only be constructed in MSpace. The drawings we have considered have been worked in a single viewport – the whole of the drawing area of the AutoCAD drawing editor. When working in a single viewport in MSpace, no viewport edges will be visible. The drawing area of the screen can be divided into a number of viewports by calling the command **VIEWPORTS** (vports) and selecting the required pattern of viewports required. When several viewports are on screen in MSpace, they can be compared with tiles on a wall – they abut each other in fixed positions and their edges are shown by lines on the screen. Only one of the viewports in MSpace will be active – i.e. allowing a drawing to be constructed within its area. The active viewport will be outlined by a double line. In order to move a viewport or change its size, the command **TILEMODE** must be called and tilemode turned off. When tilemode is off, the screen will be in **Paper Space** (PSpace), any viewport can be moved, scaled, rotated or erased but the operator can no longer work in 3D. Constructions can be carried out in 2D in any area of the screen, despite the fact that the viewport outlines will normally still be visible. When in PSpace another command **MVIEW** becomes effective.

Fig. 13.1 The **View** pull-down menu, with the **Tilemode** command

The command TILEMODE

Tilemode is set off (0) or on (1) by entering the command at the keyboard or selecting **Tilemode** from the **View** pull-down menu (Fig. 13.1):

    Command: tilemode   *right-click*

New value for TILEMODE <1>: 0 *right-click*
Command:

and tilemode is turned off and the drawing editor is in **PSpace**.

Notes

1. The usual default setting for TILEMODE is on (1). Thus when a new drawing is started, it will normally be constructed in a drawing editor set to MSpace with tilemode on;
2. With tilemode on (1), the drawing editor is in MSpace. With tilemode off (0) the drawing editor is in PSpace;
3. 3D construction can take place in MSpace with tilemode set to either 1 or 0 (on or off);
4. Drawings from earlier releases of AutoCAD can only be opened in AutoCAD 12 when tilemode is set to 1 (on);
5. Viewports are in fixed positions when in MSpace;
6. Viewports can be moved, scaled, rotated or erased when in PSpace – tilemode off (0);
7. Viewports can only be set in MSpace with tilemode on (1);
8. Viewports can be set in PSpace with the command MVIEW. The command MVIEW cannot be used in MSpace.

The command MVIEW

When **TILEMODE** is off (0), the drawing editor will be in **Paper Space** (PSpace) – the PSpace icon will be at the bottom left-hand corner (if UCSICON is ON). If previously working in MSpace with tilemode set on (1) and **TILEMODE** is then set to 0 (off), the drawing under construction disappears from the screen and will not re-appear until settings controlled by the command **MVIEW** have been made. When the command is called the command line shows:

Command: mview *right-click*
ON/OFF/Hideplot/Fit/2/3/4/Restore/<First point>:

When responses are made to these options:

1. With the response of 2, 3 or 4, further options appear allowing the operator to choose any one of the layouts indicated in the **Tiled Viewport Layout** dialogue box – Fig. 13.2;
2. If an h (Hideplot) is the response, a further option appears requesting whether Hideplot should be ON or OFF. With Hideplot on, hidden lines from the chosen viewport will not

appear when the drawing is printed or plotted. The viewport is chosen by selecting any one edge followed by a *right-click*;
3. An r (restore) will restore the previous viewport layout. The default response is ***ACTIVE** which restores the drawing to that constructed in MSpace;
4. An f (fit) asks for the first and other point for a window into which the requested number of viewports will fit;
5. On or off turns the whole of the contents of any chosen viewport on or off. The selection of the viewport is made by picking any one edge of the chosen viewport after the response off (or on if the contents have previously been turned off) has been given.

Viewports

There are two methods by which viewport layout can be set — either by calling the command **VIEWPORT** (or vport) and responding to the prompts which appear when the command is called, or be selected from the **Tiled Viewport Layout** dialogue box.

The easiest method of changing viewport configuration is to select **Layout** from the **View** pull-down menu (Fig. 13.1) followed by selecting **Tiled Vports ...** from the cascaded menu which

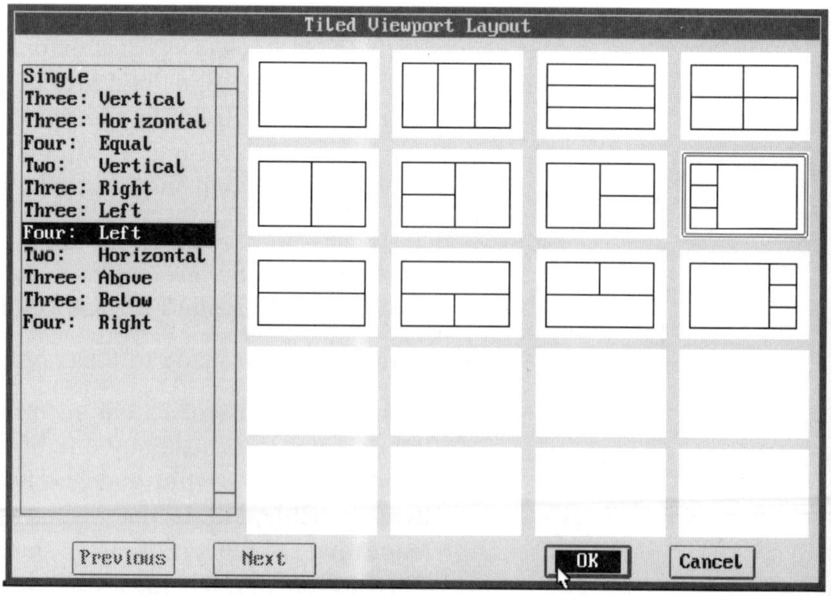

Fig. 13.2 The **Tiled Viewport Layout** dialogue box

Tilemode, MSpace, PSpace and Dview

appears. The **Tiled Viewport Layout** dialogue box appears (Fig. 13.2), from which the required viewport configuration can be selected. If an attempt to change the viewports layout is made with tilemode off, a message will appear informing the operator that the change is not possible with tilemode off.

When more than a single viewport is on screen, only one of them is active at any one time. A viewport is made active by moving the cursor into the required viewport, followed by a *right-click*.

When drawing 3D models, if one works in a 3-viewport or a 4-viewport layout, the results of each detail of construction can be seen from different viewpoints. The drawing is mostly constructed in the largest of the viewports. This allows the operator to confirm from different viewing positions whether details of the model are correctly drawn or not as the work proceeds. Two examples are given in Figs 13.3 and 13.4. Taking Fig. 13.3 first:

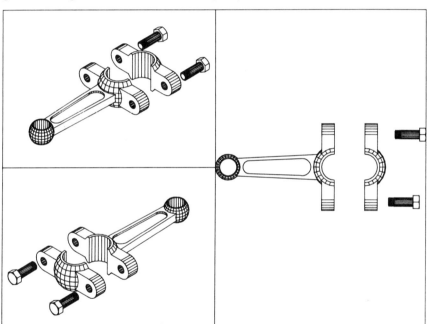

Fig. 13.3 A 3D model constructed in a 3-viewport layout

1. Check that **TILEMODE** is on (1);
2. Set viewports to **Three: Right**;
3. *Right-click* in the top left viewport to make that viewport active. Set **UCSFOLLOW** to 0 (off). Set **VPOINT** to 1,1,1;
4. *Right-click* in the bottom left viewport to make it active. Set **UCSFOLLOW** to 0 (off); set **VPOINT** to −1,−1,1;
5. *Right-click* in the right-hand viewport to make it active. Check that **UCSFOLLOW** is set on (1). Set **UCS** to **World**.

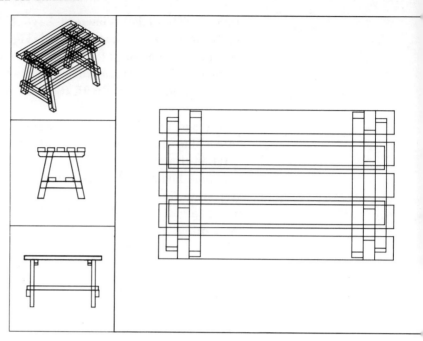

Fig. 13.4 A 3D model constructed in a 4-viewport layout

This configuration allows three different viewing points – set by **VPOINT** and **UCS WORLD** – but any change in the UCS when working in the right-hand viewport will not be followed in the other two viewports.

In Fig. 13.4, the viewports configuration is:

1. **TILEMODE** on (1);
2. Viewports set to **Four: Right**;
3. Top left viewport – **VPOINT** −1,−1,1; **UCSFOLLOW** off (0);
4. Centre left viewport – **VPOINT** −1,0,0; **UCSFOLLOW** off (0);
5. Bottom left viewport – **VPOINT** 0,−1,0; **UCSFOLLOW** off (0);
6. Right viewport – **UCSFOLLOW** on (1); **UCS WORLD**.

In each example, details of the construction are worked mainly in the right-hand large viewport. The **UCS** can be changed from the **Settings UCS Settings ...** pull-down menu as required – because **UCSFOLLOW** is set at 0 in the other viewports, changes of the UCS in the large viewport will not be reflected in the others. As work proceeds, additions will be seen in all viewports from different viewing positions, allowing the operator to check the accuracy of his/her work.

Figure 13.5 shows the 3-viewport drawing Fig. 13.3 after each viewport has been acted upon by **SOLPROF** (see page 195) and the layer on which the viewport outlines is held has been turned off, thus removing the viewport outlines.

Tilemode, MSpace, PSpace and Dview

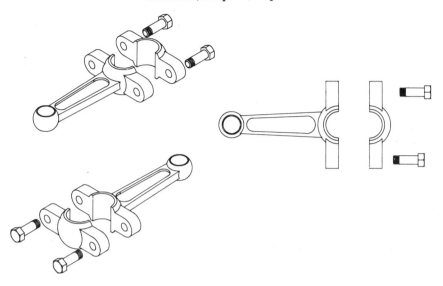

Fig. 13.5 Shows Fig. 13.3 after the action of **SOLPROF**

Orthographic projections from AME models

Orthographic projections can be obtained from AME models with the aid of the commands Vport; Tilemode; PSpace and MSpace. Figures 13.6, 13.7 and 13.8 are examples of the results of using the methods involved. The procedures follow a sequence such as:

1. In **MSPACE** construct the model drawing – Fig. 13.6;
2. **VIEWPORT** – set up the model in a 4-viewport layout;
3. **VPOINT** – top left 0,0,1; top right $-1,-1,1$; bottom left $0,-1,0$; bottom right 1,0,0;
4. **TILEMODE** 0 (off). The 4-viewports disappear from the screen and the PSpace icon appears;
5. Make a new layer VP, colour say yellow;
6. **MVIEW** – option 4 followed by f (Fit). The original screen reappears with the model drawing in each of the four viewports;
7. **MOVE** viewports to new positions – Fig. 13.7. To move a viewport select any one of its edges and the viewport with its drawing can be moved as if it were an entity;
8. Turn layer vp off. The viewport outline disappears;
9. Call **MSPACE**. Set **TILEMODE** to 1 (on). **SOLPROF** each viewport in turn;
10. **TILEMODE** to 0 (off). The screen returns to PSpace. The drawing can now be dimensioned – BUT there is no value in using **OSNAP** to determine the exact positions of the dimensioning lines – the drawing is 3D and not 2D – Fig. 13.8.

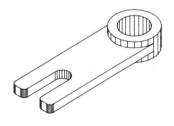

Fig. 13.6 A 3D AME model

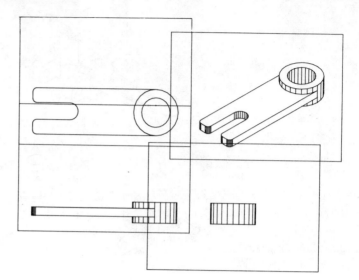

Fig. 13.7 The 3D AME model in a 4-viewport layout in **PSpace**

Another command which could be used to obtain orthographic projections from AME models is **SOLVIEW**, but discussion of this command is beyond the scope of this book.

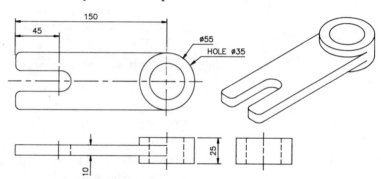

Fig. 13.8 The orthographic projection after **SOLPROF** and with dimensions added

The command DVIEW

The actions of this command can be compared with those of **VPOINT** (page 163). As with vpoint, the command dview allows 3D models to be viewed from a variety of directions. Unlike vpoint, however, the actions of **DVIEW** appear dynamically on screen under the actions of the movements of the selection device (in our case a mouse). Once an option has been chosen, a ghosted image of the model appears under the control of the mouse. A *right-click* determines the final position of the model when the operator is satisfied with the view shown by the dynamically movable ghosted model. Figure 13.9 shows the results of responses to some of the options within the command system.

Enter dview and the command line shows:

Command: dview *right-click*
**CAmera/TArget/Distance/POints/PAn/Zoom/CLip/Hide/Off/
Undo/<eXit>:**

CAmera: the model can be dynamically moved around the screen as if viewed from the position of the camera. Or: x,y,z coordinates can be entered to state the position of the camera.

TArget: action similar to that of the camera option.

Distance: when camera, target or points options have been completed, the object can be set at a distance from the camera. A slider bar appears on screen – Fig. 13.10. Although the slider bar can be moved under the action of the mouse, a fairly large number needs to be entered in response to the Distance option – see Fig. 13.9. When the Distance option is used, the model appears in perspective and the perspective icon appears at the bottom left of the screen. When this icon appears some AutoCAD commands become ineffective.

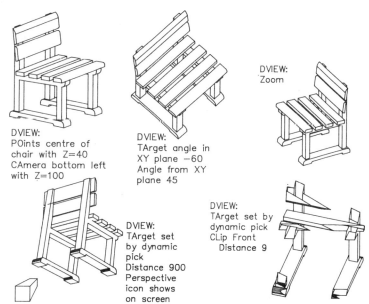

Fig. 13.9 The results of some of the **DVIEW** options

DVIEW:
POints centre of chair with Z=40
CAmera bottom left with Z=100

DVIEW:
TArget angle in XY plane −60
Angle from XY plane 45

DVIEW:
Zoom

DVIEW:
TArget set by dynamic pick
Distance 900
Perspective icon shows on screen

DVIEW:
TArget set by dynamic pick
CLip Front
Distance 9

POints: select points on target and for the camera. Usually better to enter x,y,z coordinates for the two positions.

PAn: pans the model on screen.

Zoom: the slider (Fig. 13.10) appears and the zoom scale can be set by adjusting the slider pointer under movement of the mouse.

TWist: is selected to dynamically twist the model around its centre under control of the mouse.

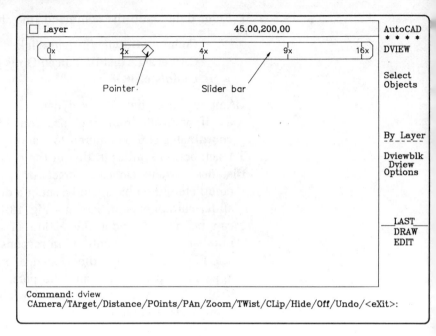

Fig. 13.10 The slider bar which appears when Distance or Zoom are selected as responses to **DVIEW** options

CLip: either the front or the back of the solid can be clipped. This allows an internal view of the object within its clipped outlines.
Undo: undoes the last dview response.
eXit: leave the dview command system.

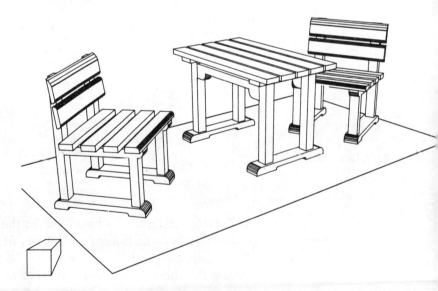

Fig. 13.11 AME solids in a **DVIEW** set by POints and Distance. Note the perspective icon

CHAPTER 14

Further exercises

Introduction

The exercises in this chapter may be suitable as part of assignments.

Exercise 1

The three drawings Figs 14.1, 14.2 and 14.3 show:

Fig. 14.1: An outline of the outer walls of a two-bedroom bungalow;
Fig. 14.2: A design for the bungalow with details of its rooms and furniture;
Fig. 14.3: The inserts from which details within the design Fig. 14.2 were added to the drawing. Examine details within the design Fig. 14.2. You will see that the design is clearly unsuitable. Make sketches freehand in pencil showing a more suitable design within the parameters set by the outline Fig. 14.1. Draw your best design with the aid of AutoCAD, using the Insert facility for

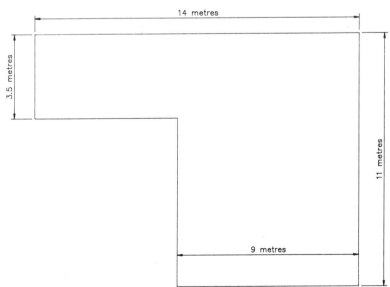

Fig. 14.1 Outline dimensions of the bungalow plan

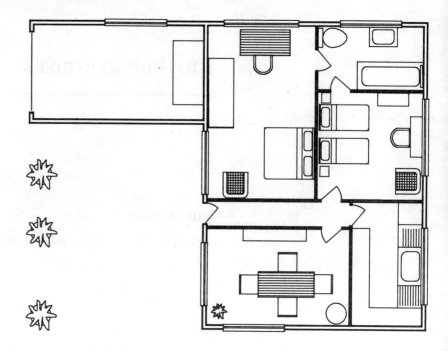

Fig. 14.2 The suggested bungalow plan

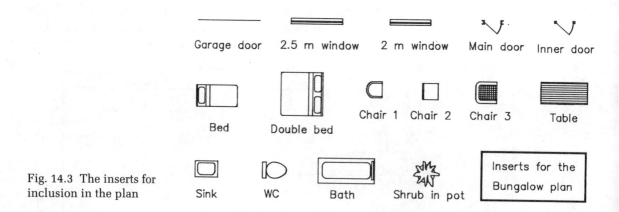

Fig. 14.3 The inserts for inclusion in the plan

inserting various parts of the drawing. Add the following to your drawing:
1. paths to the garage and main door;
2. a porch and step to the bungalow entrance.

Exercise 2

Figure 14.4 shows an assembly drawing in Third Angle orthographic projection of part of a machine. Working with the aid of AutoCAD, draw the following three views of the part, working to a scale of 1:1 on an A3 screen:

1. the given plan;
2. the given end view;
3. a sectional front view on A–A.

Include in your views a roller spindle and a device which retains the spindle within the bracket; include necessary dimensions. Your drawings should also include the following:

(a) details of the materials from which the parts should be made and details of any necessary tolerancing;
(b) details of the methods by which the parts should be manufactured.

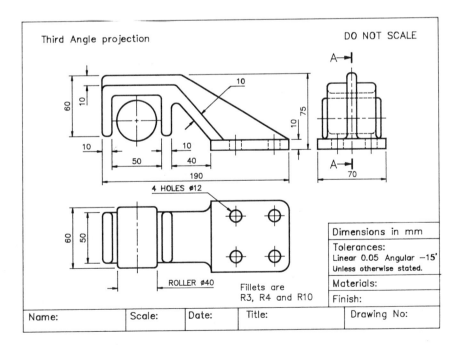

Fig. 14.4 Orthographic projection for Exercise 2

Exercise 3

Figure 14.5 is a sectional view through a Wankel engine. Working to any suitable dimensions, construct a copy of the given sectional view.

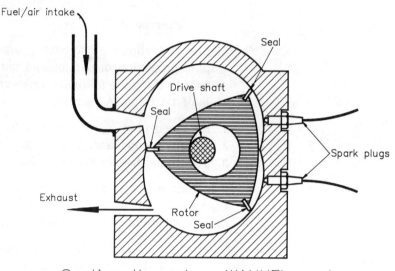

Fig. 14.5 Exercise 3

Section through a WANKEL engine

Exercise 4

Figure 14.6 is a front elevation of a house. Construct a copy of the given elevation, working to suitable dimensions. Note the following:

(a) two types of window are included in the elevation. You need to draw only one of each type and then copy it to the other three positions for each window;
(b) the hatched areas are best constructed on two layers – one to contain the hatch area outlines, the other to contain the hatching. The hatch outline layer can then be turned off.

Exercise 5

Construct an AME solid model of the device shown in Fig. 14.7.

1. Draw a pline outline of the body of the device in UCS FRONT. This can then be SOLEXTruded;
2. In UCS RIGHT construct the horizontal slot through the body. Check its position in UCS WORLD. Then SOLSUB the slot from the body;

Further exercises 217

Fig. 14.6 Exercise 4

3. In UCS WORLD, construct the vertical slot. Adjust its position in UCS FRONT. Then SOLSUB the slot from the body;
4. Construct the screw from SOLCYLs, partly in UCS WORLD, and adjust its parts in UCS FRONT before joining the SOLCYLs together with SOLUNION;
5. Construct the handle pin in UCS FRONT and adjust it for position in UCS WORLD, then join it to the screw with SOLUNION;
6. Construct the nut in UCS WORLD and adjust its position in UCS FRONT.

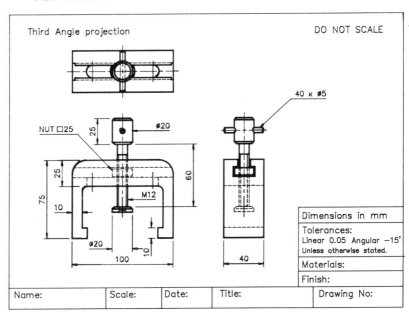

Fig. 14.7 Exercise 5

Exercise 6

Construct an AME solid model of the component shown in Fig. 14.8.

1. In UCS FRONT, construct a pline outline of the front upper part of the component. With SOLEXT, extrude it to a height of 10;
2. In UCS WORLD copy the front extrusion to form the back;
3. In UCS WORLD construct an extrusion of height 10 of the base of the upper part of the component. Check its position in UCS FRONT. SOLUNION the three parts;
4. SOLFILL the curves at the base of the upper part of the component;
5. The vertical 20-mm-thick support piece can be constructed in a similar manner in UCS RIGHT, checking its position in UCS WORLD, before forming a union with the upper union;
6. The collars and holes are formed from SOLCYLs.

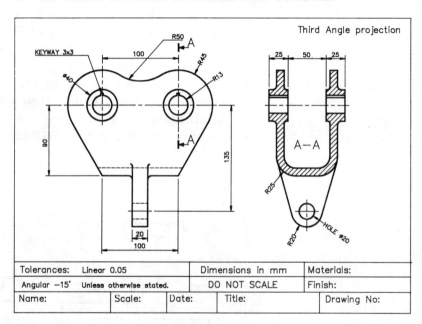

Fig. 14.8 Exercise 6

Exercise 7

Figure 14.9 is a sectional view through a double glazing window fitment. Copy the given drawing working in an AutoCAD drawing editor configured to A3 size sheet. Work to any suitable dimensions.

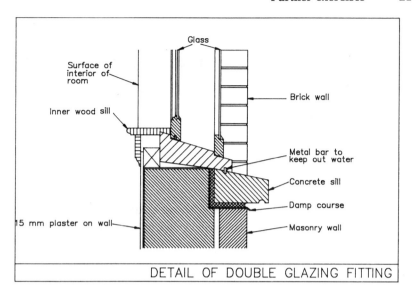

Fig. 14.9 Exercise 7

APPENDIX A

Orthographic projection

In this form of projection, it is imagined that a three-dimensional (3D) object being drawn is placed in one angle of two crossing planes, one of which is horizontal (the **Horizontal Plane** or **HP**), the other vertical (the **Vertical Plane** or **VP**). The angle between an HP and a VP is a right angle (geometrically described as being orthogonal to each other – hence the name orthographic). This theory is explained in a series of 14 drawings in Figs A.1 to A.8. These drawings show:

Drawing 1

The two orthogonal planes within one angle of which the 3D object being drawn will be placed. The direction in which it will be looked at to obtain orthographic views are shown.

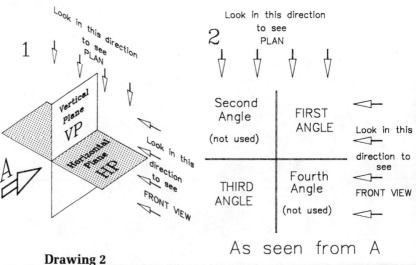

Fig. A.1 Orthographic projection, drawings 1 and 2

Drawing 2

Two planes as seen from the front. Each of the dihedral angles (angles made by planes) is named as either **FIRST**, **SECOND**,

Appendix A

THIRD or FOURTH ANGLE. For the purposes of orthographic projection only the **First** and **Third** angle are used.

Drawing 3

Place the 3D object to be drawn in the **FIRST ANGLE** quadrant. Draw the **FRONT VIEW** (or front elevation) on the **VP**. Draw the **PLAN** on the **HP**.

Drawing 4

The 3D object has been removed from the First Angle quadrant.

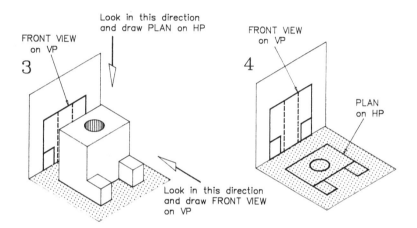

Fig. A.2 Orthographic projection, drawings 3 and 4

Drawing 5

Rotate the **HP** with its plan drawing through 90 degrees until:

Drawing 6

The two views – Front View and Plan – are in line as if on a 2D sheet of paper. These two views are said to be in **FIRST ANGLE PROJECTION**.

Drawing 7

Place the 3D object in the **THIRD** dihedral angle. View in the same directions as for **First Angle**. Draw the **FRONT VIEW** on the **HP** and draw the **PLAN** on the **VP**.

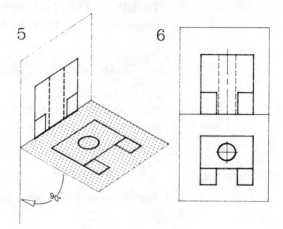

Fig. A.3 Orthographic projection, drawings 5 and 6

Drawing 8

The 3D object has been removed from the Third Angle quadrant. Rotate the **HP** with its plan drawing through 90 degrees.

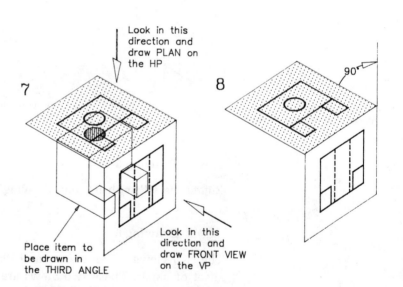

Fig. A.4 Orthographic projection, drawings 7 and 8

Drawing 9

The two views – Front View and Plan – are in line as if on a 2D sheet of paper. These two views are said to be in **THIRD ANGLE PROJECTION**.

Appendix A   223

Drawing 10

Using the same theoretical method another view – an **END VIEW** – can be drawn on a second **VP** placed to the right of the 3D object.

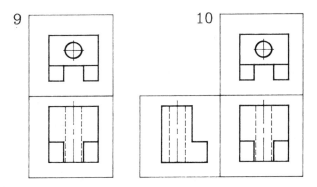

Fig. A.5 Orthographic projection, drawings 9 and 10

Drawing 11

By the addition of further imaginary vertical or horizontal planes, as many as six views can be constructed. Such a complete set of views is shown in this drawing.

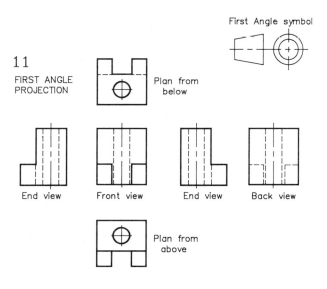

Fig. A.6 Six views in First Angle projection, drawing 11

Drawing 12

The six views in the Third Angle.

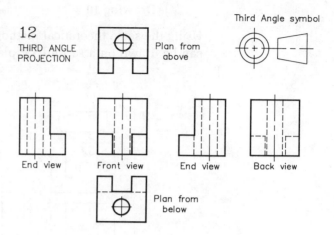

Fig. A.7 Six views in Third Angle projection, drawing 12

Notes on First and Third Angles

1. Although as many as six views — indeed more if thought necessary — can be constructed in orthographic projection, normally one, two or three views are all that are necessary to completely describe a 3D object. For example an object made from thin sheet material would probably need only one view; other more complex objects would need more views;
2. From the drawings above, it will be seen that, although First Angle projections are similar to Third Angle, there are differences in the positioning of views and plans in relation to each other;
3. In **First** Angle a plan as seen from above is **below** the front view;
4. In **Third** Angle a plan seen from above is **above** the front view;
5. In **First** Angle an end view is placed on the **opposite** side of the front view to that side from which the object is viewed;
6. In **Third** Angle an end view is placed on the **same** side of the front view as the side from which the object is viewed;
7. In **First** Angle both end views and plans face **away** from the front view;
8. In **Third** Angle both end views and plans face **inwards** towards the front view.

Sectional views

To determine the shape of the inner parts of a 3D object, it is imagined that a plane (vertical or horizontal) is passed right

Appendix A   225

through the object and what is seen is then drawn on the plane. The theory of sectional planes is illustrated in Fig. A.8.

When a sectional view is included in a set of orthographic projections it should be drawn to conform to a set of rules. Some of these are indicated in the plan and sectional front view of Fig. A.9. The rules include:

1. All the detail seen behind the cutting plane are included with the section;
2. Features such as webs, ribs, bolts, nuts and other cylindrical parts are shown by outside views within the sectional view;
3. The section cutting line and the direction of viewing to see that cut surface is usually included;
4. Sectional views are normally labelled.

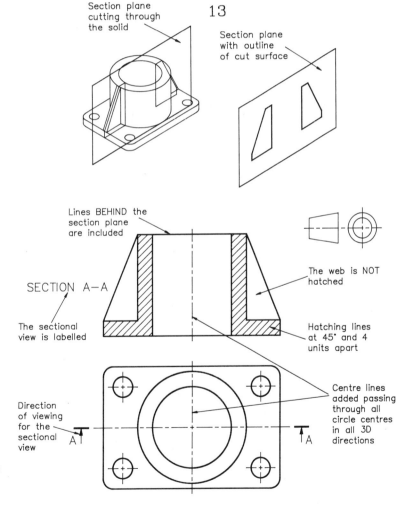

Fig. A.8 The theory of section cutting planes, drawing 13

Fig. A.9 A sectional front view and plan in First Angle orthographic projection

Types of line in technical drawings

In general several types of line will be found in technical drawings. The more important of these are shown in Fig. A.10. In fact, the types of line shown in this illustration are peculiar to engineering drawing. Other types of drawings such as building drawings will have somewhat different lines. Figure A.10 is, however, a good guide for many forms of technical drawing.

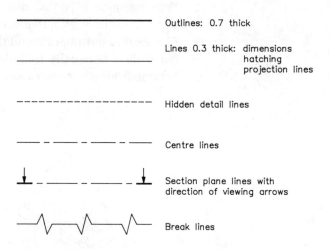

Fig. A.10 Types of line in technical drawings

APPENDIX B

MS-DOS

Introduction

In this book we are primarily concerned with AutoCAD 386 Release 12 – i.e. with the AutoCAD software installed in a 386 or 486 computer running under MS-DOS. It must be emphasized, however, that its contents are suitable for AutoCAD 12 installed in other types of computer. In this appendix some information regarding commands in MS-DOS as they affect the use of AutoCAD are briefly described. Although AutoCAD 386 Release 12 will run with any MS-DOS system from version 3.3 onwards, the best AutoCAD 12 performance will be with MS-DOS version 5.0 or higher. MS-DOS is an abbreviation for Microsoft Disk Operating System. Microsoft is an American software company. MS-DOS is a complex program and this chapter gives only a very brief summary of a few of the numerous commands available with the program.

Start-up

When a PC (Personal Computer) running under MS-DOS is switched on a self-test is run – memory, system circuits, disk drives and other peripherals are checked. When satisfied that these are OK, some of the MS-DOS files are loaded into memory. At this stage only those commands most frequently used are loaded. Commands not frequently used are loaded as and when they are required.

The PC may have several drives – see Fig. B.1. AutoCAD 12 files must be loaded on a hard disk, so in the example given in Fig. B.1 the files will be on drive C:\>. PCs may have more drives than those shown in Fig. B.1 – up to as many as 26 are possible working under MS-DOS, although it is more usual to have five or six – e.g. floppy disk drives A:\ and B:\, hard disk drives C:\, D:\ and E:\ say perhaps and a CD-ROM drive F:\.

228 AutoCAD Release 12 for students

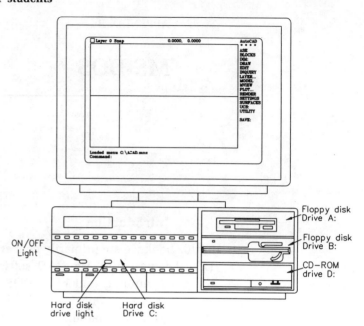

Fig. B.1 A typical PC set-up such as would be used at an AutoCAD 12 workstation

When the start-up test procedure is completed, a series of statements may appear on screen, usually finishing with the MS-DOS command line prompt **C:\>**. MS-DOS commands are entered at this prompt. With MS-DOS version 5.0, it is possible to have the start-up showing a graphics screen from which commands etc. can be selected with the aid of a mouse. Here we will only be concerned with commands entered from the keyboard.

When a PC is switched on from the power switch, this is said to be a *cold start*. When the computer has been running for some time, if a fault occurs a *warm start* can be made by pressing the three keys Ctrl/Alt/Del. It is advisable to avoid too many cold starts when using a PC, because repeated cold starts may damage the chips in the system circuits. Warm starts do not carry this risk. Avoid warm starts while working in AutoCAD 12 if possible – although your current drawing will usually be saved in a file named *AUTO.SV$*, it is still possible to lose some of your construction. A good practice in any case is to save your work at regular intervals – say every 15 minutes or so. Then if anything goes wrong – e.g. a fault in the power supply – at least most of your work will have been saved.

Some MS-DOS commands

Either capital or lower case letters can be used for any of the MS-DOS commands.

To change from drive C:\> to drive A:\>

    C:\> a: *Return*
    A:\>

and drive A:\> becomes the current drive.

To list the files on a disk

C:\> dir   *Return*
Volume in drive C is hard-disk
Volume serial Number is 18A9–4186
Directory of C:

DOS	<DIR>		20/10/91	17:18
ACAD12	<DIR>		19/08/92	11.25
COMMAND	COM	175	19/06/92	9.35
CONFIG	SYS	307	19/06/92	10:53
AUTOEXEC	BAT	332	19/06/92	10.59
W	BAT	8	05/08/92	8:47

        6 file(s)   **822 bytes**
        12429526 bytes free

Taking examples from the above directory:

Volume is the name that is given to the hard disk by the operator;
Volume serial number is automatically given to a disk by MS-DOS version 5.0 (or later);
ACAD12 is the name of the directory holding the AutoCAD 12 files;
<DIR> shows that this is the name of a directory;
19/08/92 is the directory was made on the 19th August 1992;
11.25 is the time of the day (24-hour clock) when the directory or file was saved to disk;
AUTOEXEC BAT – AUTOEXEC is the name of a file with **BAT** being the file extension
332 against the filename and its extension is the number of bytes in that file.

To list files in a directory

    C:\> dir acad12   *Return*

Will give a list of all sub-directories and files in the directory acad12.

Directories, sub-directories and files

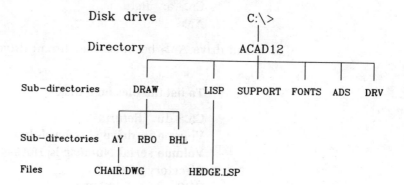

Fig. B.2 The MS-DOS directory and files structure

Figure B.2 shows the structure of the MS-DOS disk filing hierarchy. In the examples given in Fig. B.2:

C:\> is the name of the drive holding the directories and files;

ACAD12 is the name of the directory holding the AutoCAD 12 system files;

DRAW is a sub-directory of ACAD12, which itself has three sub-directories:

AY is the sub-directory of DRAW in which my drawing files (those constructed by A. Yarwood) are held;

RBO is the sub-directory of DRAW in which the drawing files of the operator whose initials are RBO are held;

BHL is the sub-directory of DRAW in which the drawing files of the operator whose initials are BHL are held;

LISP is the sub-directory of ACAD12 holding all the AutoLISP files;

DRV is the sub-directory of ACAD12 holding all the monitor, digitizer (mouse etc.) and plotter driver files.

SUPPORT is the sub-directory holding ACAD12 support files;

ADS is the sub-directory holding AutoCAD ADS (AutoCAD Development System) files;

FONTS is the sub-directory holding the AutoCAD font files.

To make a directory

C:\> md (or mkdir) acad12
C:\>

and the directory **ACAD12** is formed on drive **C:\>**.

To change directories

C:\\> cd acad12
C:\\>ACAD12

To make a sub-directory in ACAD12

C:\\> ACAD 12 md draw
C:\\> ACAD12

Filenames

Filenames usually have a file extension. The file **HEDGE** in sub-directory **ACAD12\\LISP** will have a filename extension *.lsp*. Thus its full name is:

acad12\\lisp\\hedge.lsp

Note the back-slashes between directory and sub-directories (or filenames) and the fullstop after the filename and before its extension. These are obligatory.

Filename extensions

A number of filename extensions are used in AutoCAD 12 files. Those in most common use are:

.dwg extension for an AutoCAD drawing file;
.bak extension for a backup file. If an AutoCAD drawing file is saved a second (or more) times, a backup file (extension *.bak*) will form automatically on disk;
.exe (or *.com*) files which execute a software program. An example is *acad.exe* held in the directory *acad12*. When AutoCAD is called, the *acad.exe* loads into memory and the program commences;
.cfg a configuration file;
.sld an AutoCAD slide file;
.lsp an AutoCAD AutoLISP file;
.exp among other files – AutoCAD ADI driver files.

To copy a file from one directory to another

C: copy acad12\\draw\\ay\\chair.dwg acad12\\draw\\rbo

and the file *chair.dwg* is copied from the *ay* sub-directory into the *rbo* sub-directory.

To copy a number of files between directories

C:\> copy\acad12\ay\draw*.*acad12\draw\rbo

will copy all files in the directory c:\acad12\draw\ay to acad12\draw\rbo.

To rename a file

C:\> rename acad12\draw\ay\chair.dwg acad12\draw\ay\armchair.dwg

To erase a file

C:\> erase (or del) acad12\draw\ay\chair.dwg

and the file *chair.dwg* is deleted from the disk.

Note: Take great care when erasing. If a backup file (extension .bak) has replaced the erased file, it will be possible to rename it to the filename it is backing up. Otherwise a file may be lost. MS-DOS version 5.0 does, however, have an **UNDELETE** command, which may save such a situation.

To send the contents of a file to a printer

C:\> type readme.txt>prn
or
C:\> readme.doc>prn

Note that this form of usage is not of any use for most files from software packages because they will probably be written in machine code. Only files such as those ending in .txt or .doc or other text files are suitable for this usage.

APPENDIX C

Plotting

In AutoCAD 12 both plotters and printers work from the same set-up, which is controlled by either entering the command **PLOT** at the command line, selecting **PLOT ...** from the AutoCAD main on-screen menu or selecting **Plot** from the **Draw** pull-down menu. No matter which of these actions is taken, the **Plot Configuration** dialogue box appears on screen – Fig. C.1. Settings are made from this dialogue box. Several of the boxes within the dialogue box have their own dialogue boxes. The various settings required for plotting or printing are very easily made in the dialogue box(es). Because of this only a bare outline of the settings is given below. The boxes in the dialogue box(es) containing the configuration options in the dialogue box are:

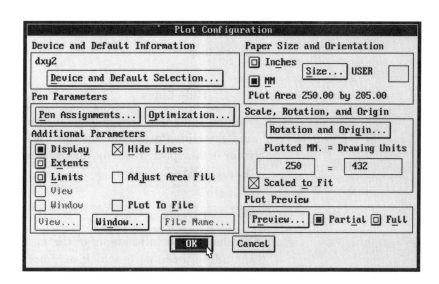

Fig. C.1 The **Plot Configuration** dialogue box

Device and Default Selection ...

A number of plotters and/or printers can be configured in AutoCAD 12. A *left-click* in this box displays a list of the configured plotters, from which a selection can be made.

Pen Assignments ...

A *right-click* in this box brings up a dialogue box in which changes to pen assignments can be made for those plotters which have multiple pens.

Size ...

A *left-click* brings up a dialogue box **Paper Size** with the previously set paper sizes from which the desired plot paper size can be selected.

Rotation and Origin ...

Calls a dialogue box in which settings of an origin and/or a plot rotation can be made.

Scaled to Fit

If checked – diagonals in box – the drawing being plotted will be scaled to the size set in the **Paper Size** dialogue box.

Display

Check to plot the whole of the area displayed on screen.

Hide Lines

Hidden lines will be removed in the plot. Note that if in **PSpace**, and the **Hideplot** option has been selected to remove hidden lines from a viewport, there is no need to select **Hide Lines** from the dialogue box.

Extents

If selected the full extents of the current drawing will be plotted, including those parts which may not be visible on screen.

Limits

Plots the area within the coordinates defined by **Limits**.

Plot to File

The plot will be sent to a file, the name of which is entered in the **File Name** dialogue box. The plot will be saved in a file with an extension *.plt*. To plot from such a file – at the MS-DOS command prompt:

    C:\> type ay.plt>com2   *Return*

Note: The ports to which plotters or printers can be attached are:

    LPT1 (LPT2, LPT3, etc.) – parallel ports;
    COM1 (COM2, COM3, etc.) – serial ports.

Plot Preview

Allows a partial or a full preview of the plot on screen.

Ctrl+C

At any time, plotting can be stopped by pressing these two keys. The plotter may take some time to cease functioning depending upon the size of the plotter's buffers.

Index

*.dwg 10
*.shx 10
*.sld 10
386 2
386 chip 17
3D
  coordinates 160
  objects 161
  surfaces commands 160
3dface 161, 162
3dobjects 161

A3 sheet size 24
abbreviations 12
absolute coordinates 12
acad.dwg 18
acad.pgp file 12, 23
accuracy in drawings 24
Add 52
adding text to drawings 42
Advanced Modelling Extension
    see AME
advantages of CAD 1
All 52
alternative dimensions 115
AME 160, 172
  Boolean operators 174
  orthographic projection 209
  primitives 172
  solid models 177, 182
angles 26
anti-clockwise 26
Arc 31
Array 87

Assist pull-down menu 44
basic rules 1
hatch files 3
Bhatch 93, 94
Block 140
Boolean operators see AME
BOX 52
Break 68
building drawing 145
  symbols 145

calling command 22
cascading menus 9
Chamfer 90
Change 80
check acad.dwg file 19
Circle 33
circuit diagram drawing 135
circuit symbols
  electrical 143
  electronic 143
cold start 227
command line 5
  prompts 6
commands
  3dface 161
  3dobjects 161
  Arc 31
  Array 87
  Bhatch 94
  Block 140
  Break 68
  Chamfer 90

Index

Change 80
Circle 33
Copy 60
Ddinsert 137
Dim 110
Dline 28
Dview 210
Edgesurf 161
Elevation 170
Ellipse 34, 151
Erase 57
Explode 140
Extend 70
Fill 38
Fillet 91
Hide 163
Isoplane 150
Line 25
Mirror 85
Move 51
Mview 205
Offset 92
Pedit 71
Pline 37
Polygon 36
Revsurf 161, 168
Rulesurf 161, 167
Scale 63
Sketch 31
Snap 149
Solbox 173
Solcham 178
Solcut 194
Solcyl 174
Solext 180
Solfill 178
Solprof 195
Solrev 180
Solsect 196
Soltorus 174
Solwedge 173
Stretch 65
Style 39
Tabsurf 168
Tilemode 204
Trim 66
UCS 190, 206

Vpoint 163
Wblock 135
Zoom 53
configured angles 26
coordinates
  x,y 24
  x,y,z 160
Coords 14, 16
Copy 60
copying files 232
CPolygon 52
Create New Drawing 8
creating a new drawing 21
Crossing 52
Ctrl+C 17
current options 23

Ddinsert 137
ddmodes 18
dialogue boxes 8, 13
dialogue box title bar 10
digitizing tablet 3
Dim 110
Dimension Style 106
dimensioning rules 116
dimensions 107
  alternative units 115
  arrow types 112
directories 230
disk drive names 11
Dline 28
Drawing Aids 15, 152
drawing editor 5
drawing sheet layout 123
Draw pull-down menu 34, 161
Dtext 45
Dview 204, 210

Edgesurf 161, 166
Elevation 160, 170
Ellipse 34, 151
Enter key 3
entities 24, 56
Entity Creation Modes 9, 41
Erase 57
erasing files 232
Explode 140

Index

Extend 70

Fence 52
file edit box 11
file pattern box 10
filenames 231
   extensions 10, 231
Fill 38
Fillet 91
first and third angles 224
first angle projection 124
fonts 39
function key calls 17

Grid 15
Grips 119

hatching 93
Hide 163
horizontal plane see HP
HP 220

IBM 2
isometric drawing 153
Isoplane 150

joystick 3

keyboard 3

Last 52
Layer Control 76
layer options 76
layers 76
left-click 3, 22
libraries of blocks 141
limits 14
Line 25
line, types of 226
Linetypes 78
list box 11
listing files 229
logic gate circuit
   drawing 144
   symbols 146

major axis of ellipse 35

minor axis of ellipse 35
Mirror 85
Mirrtext 87
modifying dimensions 113
modifying drawings 51
Modify menu commands 55
mouse 2
mouse buttons 3
   use of 25
Move 51, 59
MS-DOS 2, 228
   5.0 17
   commands 227
   prompt 3
MSpace 204
Multiple 53
Mview 205

New Drawing Name 7
New file 7

Object Snap 44
object snaps 43
oblique drawing 157
Offset 92
on-screen menus 5, 13
Open Drawing 6
Open file 6
Ortho 16
orthographic projection 123, 220

PC 2
Pedit 71
pen-up/down 31
pick-box 43
pictorial drawing 149
planometric drawing 159
Pline 37
pline arcs 39
plot preview 235
plotting 233
pneumatic circuit
   drawing 144
   symbols 145
polar arrays 89
Polygon 36
polyline 37